CANADA'S PAST AND FUTURE IN LATIN AMERICA

Edited by Pablo Heidrich and Laura Macdonald

Many historians and political scientists argue that ties between Canada and Latin America have been weak and intermittent because of lack of mutual interest and common objectives. Has this record of diverging paths changed as Canada has attempted to expand its economic and diplomatic ties with the region? Has Canada become an imperialist power?

Canada's Past and Future in Latin America investigates the historical origins of and more recent developments in Canadian foreign policy in the region. It offers a detailed evaluation of the Harper and Trudeau governments' approaches to Latin America, touching on political diplomacy, bilateral development cooperation, and civil society initiatives.

Leading scholars of Canada–Latin America relations offer insights from unique perspectives on a range of issues, such as the impact of Canadian mining investment, security relations, democracy promotion, and the changing nature of Latin American migration to Canada. Drawing on archival research, field interviews, and primary sources, *Canada's Past and Future in Latin America* advances our understanding of Canadian engagement with the region and evaluates options for building stronger ties in the future.

PABLO HEIDRICH is an associate professor in the Global and International Studies program at Carleton University.

LAURA MACDONALD is a professor in the Department of Political Science and Institute of Political Economy at Carleton University.

Canada's Past and Future in Latin America

EDITED BY PABLO HEIDRICH
AND LAURA MACDONALD

UNIVERSITY OF TORONTO PRESS
Toronto Buffalo London

© University of Toronto Press 2022
Toronto Buffalo London
utorontopress.com

ISBN 978-1-4875-4042-5 (cloth)   ISBN 978-1-4875-4045-6 (EPUB)
ISBN 978-1-4875-4043-2 (paper)   ISBN 978-1-4875-4044-9 (PDF)

Library and Archives Canada Cataloguing in Publication

Title: Canada's past and future in Latin America / edited by Pablo Heidrich and Laura
   Macdonald.
Names: Heidrich, Pablo, editor. | Macdonald, Laura, 1960– editor.
Description: Includes bibliographical references and index.
Identifiers: Canadiana (print) 20210379073 | Canadiana (ebook)
   20210379154 | ISBN 9781487540425 (cloth) | ISBN 9781487540432 (paper) |
   ISBN 9781487540456 (EPUB) | ISBN 9781487540449 (PDF)
Subjects: LCSH: Canada – Foreign relations – Latin America. |
   LCSH: Latin America – Foreign relations – Canada.
Classification: LCC FC244.L3 C36 2022 | DDC 327.7108 – dc23

This book has been published with the help of a grant from the Federation for the
Humanities and Social Sciences, through the Awards to Scholarly Publications Program,
using funds provided by the Social Sciences and Humanities Research Council of
Canada.

We wish to acknowledge the land on which the University of Toronto Press
operates. This land is the traditional territory of the Wendat, the Anishnaabeg, the
Haudenosaunee, the Métis, and the Mississaugas of the Credit First Nation.

University of Toronto Press acknowledges the financial support of the Government of
Canada, the Canada Council for the Arts, and the Ontario Arts Council, an agency of
the Government of Ontario, for its publishing activities.

Canada Council   Conseil des Arts
for the Arts   du Canada

ONTARIO ARTS COUNCIL
CONSEIL DES ARTS DE L'ONTARIO
an Ontario government agency
un organisme du gouvernement de l'Ontario

Funded by the   Financé par le
Government   gouvernement
of Canada   du Canada

Canadä

Contents

Preface

What is Canada's future in the Americas? Despite the fact that Canada has been searching for ways to diversify its international relations, political alliances, and trading partners over the last several decades, and some Latin American states are among the most dynamic developing nations and emerging markets in the world, concrete sign of improvements in the relationship are scarce. This is not a new problem. Historians and political scientists have long argued that ties between Canada and Latin America were weak and intermittent for most of the last century. The Mulroney government's decision to occupy Canada's long-empty seat in the assembly of the Organization of American States in 1990 is often seen as a turning point in Canada's relations with the region. Even after that point, however, the relationship has frequently been described as fluctuating and non-committal, reflecting weak historical precedents but also the absence of strong economic, social, political, and cultural ties. As well, successive Canadian governments lacked a sustained commitment to promoting a deeper relationship with their Latin American counterparts. In part this reflects an instinctive long-standing preference on the part of Canadian governments to prioritize their relationships with the United States, their largest market and traditional ally, and with European counterparts that are viewed as Canadian equals. Governments like that of Paul Martin that wished to signal Canada's commitment to help the poorest and most marginalized parts of the world tended to turn to Africa instead (Cameron 2007).

Stephen Harper's Conservative Party government (2006–15) did prioritize Latin America in its foreign policy with the development of an "Americas Strategy." This strategy, announced by Harper in Santiago, Chile, in 2007, did lead to higher levels of diplomatic engagement with the region and flows of development assistance and more frequent exchanges between Canadian politicians and bureaucrats and representatives from the region. However, the effectiveness of these commitments was undermined by the dominance of economic motivations and the government's failure to sustain a commitment to supporting democratic development and human rights commitments with the same intensity as it pursued commercial advantage. Despite the Trudeau government's rhetorical commitment to liberal internationalism, we argue that in practice there has been more continuity than change in this new government's approach to the region since 2015.

Thus, the question remains: Why has it been so difficult for Canada to develop a closer relationship with Latin America? At a most essential level, one could also ask: Is Canada really a country of the Americas? Or is it inevitable that historical precedents and allegiances, economic interests and the predominance of the United States in the worldview of Canadian policymakers will overwhelm any professed interest in closer ties with these countries? The authors in this collection do not have a single answer to these questions, but we hope their combined perspectives and analyses will lead to a deeper understanding of the barriers to Canadian–Latin American collaboration, as well as the potential benefits of that relationship. In this volume we focus almost exclusively on Canada's ties with Latin American countries – in which category we include Haiti – but do not deal in detail with other countries of the Caribbean and South America that have a distinct historical, cultural, and political identity. The logic for that is to gain focus on how Canada has approached a region where nations share a significant level of common history, relatively similar patterns of political and economic development, as well as increasing efforts to develop pan-regional responses to the rest of the world.

Overview of the Collection

The book begins with a series of chapters that set the historical background for the study of Canada–Latin America relations. The introductory essay by Heidrich and Macdonald provides an overview of Canada's involvement in the Latin American region. First, we cover the historical background of how Canada has gotten to where it stands nowadays in its relations with this region, then critically assess the main theoretical writings done about Canadian foreign policy towards Latin America, to finally position the contribution of this volume for the analysis of the last couple of decades of bilateral relations.

The next two chapters take a historical perspective on the character of Canada–Latin America relations and how they have evolved over time. Chapter 2, by Canadian historian Asa McKercher, uses the concept of "mental maps" to elucidate how Canadian foreign policymakers have viewed their country's relationship with the rest of the Americas (apart from the United States) in geospatial terms. He argues that geographic and cultural location have been key elements of the way in which many Canadians have seen themselves in the world and helps to explain why, historically, foreign policymakers tended to ignore the Western Hemisphere. The chapter draws on a variety of historical government, press, and academic sources to offer a critical analysis of elite Canadian opinion regarding Canada's position within the Western Hemisphere. McKercher demonstrates the underlying power dynamics that lead foreign-policy decision-makers to view themselves as located in the trans-Atlantic core of the world system, and politically, economically, culturally, and geographically distant from peripheral nations in the Americas. While Latin America was largely absent from or extremely distant in the mental maps of most Canadian foreign policymakers, there were some adventurous Canadian politicians or bureaucrats who attempted to redraw the map in the eyes of Canadians to include the region in a more prominent form. However, it was more frequently businessmen (and they were mostly men) who could see the importance of the region to the economic interests of themselves and other Canadians. The importance of economic interests in breaking through entrenched assumptions about the region is a recurring theme, but, as discussed in other chapters in the volume, it did not necessarily lead to more egalitarian views of the rights of Latin Americans.

McKercher argues, in fact, that given the cognitive perspectives that shaped dominant perspectives, it was left to Canadian non-state actors to lead the way in reconceptualizing Canada's role in the world and promote alternative perspectives and worldviews. This argument is supported by the work of Canadian historian and activist John Foster, who in Chapter 3 presents an insider's perspective on the work of the Latin American Working Group (LAWG), a pioneering research and solidarity organization that was formed in Toronto, Ontario, in 1966 in direct response to the 1965 US invasion of the Dominican Republic. Its aim was to educate Canadians on the socio-economic realities of the people of Latin America and to support their struggles for social justice. LAWG focused its research on Canada's role in the region with special emphasis on Canadian aid policies and the role of multinational corporations. LAWG worked in solidarity with the Canadian churches, trade unions, and a wide variety of other non-governmental and grassroots organizations to raise public awareness and to lobby for changes in Canadian government policies. The chapter is based on the work of former members of the LAWG collective to document their history and preserve historical records of their work. It provides a fascinating counterpart to McKercher's portrait of elite opinion, since it shows how civil society actors, guided by ethical and ideological imperatives, contested the dominant interests and opinions that shaped official policy. While never as successful as the members hoped, they did pave the way for stronger Canada–Latin America relations based on people-to-people ties. These themes are explored in the chapters in this volume by Deonandan and Schmuland and Moore.

The next group of chapters focuses on several aspects of Canadian foreign and security policies. Chapter 4 by Colombian scholar Federmán Rodríguez provides an overview of the evolution of Canadian security and defence policies in the Americas. Like McKercher, Rodríguez emphasizes the importance of the analytical lenses that affect Canadian decision-makers' perspectives. Following an insight from former Canadian diplomat Alan Gotlieb, he points to the split personality that shapes Canadians' approach to the world (and in this case specifically to the Americas). On the one hand, like other actors, Canadians sometimes adopt hard-headed realist approaches that focus on the promotion of national interest, while at others they seem to embrace a "liberal, broad, and even romantic approach to security." He argues that during the Cold War and the early years after the Cold War Canada managed to achieve a balance between these liberal values and strategic interests through what Kim Richard Nossal refers to as "liberal realism" (2007). However, in later years, he maintains, Canadian "economic security interests" have tended to undermine liberal policies towards the region, as the involvement of Canadian-owned mining companies became more intense (a theme which is explored in the final section of the book). Based on this analysis, Rodríguez reviews some of the controversies surrounding the Harper government's Americas Strategy. While some initiatives in the realm of policing assistance and security reform seem to continue the liberal orientation of earlier Canadian governments, these were, he argues, undermined by the government's insufficient attention to human rights, its ideological bias towards right-leaning government, and its prioritization of the economic interests of Canadian corporations.

The case of Haiti is explored in greater depth in Chapter 5 by Canadian political scientist Yasmine Shamsie. One of the most central elements of the liberal agenda discussed by Rodríguez is that of democracy promotion in the hemisphere. Yet how to cultivate truly democratic political systems is a controversial topic, particularly in the context of the political instability and economic inequality that characterize many

countries in the Americas. Shamsie provides insight into these complexities in a case study examining Canadian response to the 2015 elections and their re-run in 2016. Haiti is one of the places in the region
where Canada has been most active both in providing development
assistance and in its diplomatic involvement in promoting certain economic and political approaches. Haitians' perceptions of Canada's role
have been shaped, Shamsie argues, by the country's failure to come to
the support of democratically elected President Aristide when he was
toppled in a 2004 coup. Since then, Canada has been actively involved
in traditional democracy-promotion activities, such as providing support for elections, the police, judicial reform, and reforms of the penal
system, as well as neo-liberal economic reforms. This liberal approach
was tested, Shamsie argues, by the circumstances of the 2015 elections.
Differences emerged between foreign and domestic groups of observers about the adequacy of those elections: While foreign observers like
the EU and OAS viewed them as flawed but adequate, local observers
were much more critical and condemned the exercise as fundamentally undemocratic. The election results were finally annulled, and the
elections were run again the following year, against the objections of
Canada and other international actors. Shamsie argues that the Canadian position reflected the insufficient respect the northern country
placed on the opinions of Haitians. She argues that given the newness
of the Haitian democratic system, the lack of a sense of social trust and
weakness of institutions, this type of crisis could be seen as a valuable
opportunity for institutional learning and improvement that could help
build public confidence in the long-term functioning of the system. In
response, she calls for Canada to adopt a more flexible approach and
to avoid excessive involvement in the domestic politics of the country.
This approach would involve a recognition of the deep structural barriers to democratization in Haiti, given its long legacy of social and
economic injustices, which means democracy cannot be achieved easily
and smoothly but is the subject of struggle between elites and the more
marginalized sectors of the population.

While the chapters by Rodríguez and Shamsie focus primarily on
state-to-state relations, another phenomenon of increasing relevance
is the movement of Latin Americans to Canada, studied in Chapter 6
by Laura Macdonald and Christina Gabriel. This movement shows the
importance of personal histories and life stories and reflects many of
the complexities of the broader relationship. Literature on Latin American migration to Canada has primarily focused on so-called waves
of migrants from south to north and has tended to assume that these

movements are permanent and that individuals are eventually assimi-
lated into Canadian society. Gabriel and Macdonald show in their
chapter, however, how changes in both the Latin American region as
well as in Canadian policies have led to a new phase in the movements
of people between the two sites characterized by greater prevalence of
temporary migration and by more sustained and continued cultural,
economic, and social exchanges between "home" and "host" locations.
These changes, they contend, mean that the concept of "migrant transnationalism" better reflects migrants' experiences than the outdated
concept of "waves" of migrants. Their chapter reviews the history of
Latin American migration to Canada and describes recent changes,
such as the expansion of the temporary foreign worker programme
(TFWP) and the inclusion of chapters in trade agreements since NAFTA
that permit temporary mobility of skilled migrants. At the same time,
bordering practices have also intensified in recent years that have made
some migrants' journeys to Canada more difficult.

The final set of chapters in the collection shift to a focus on economic
relations between Canada and the Americas, with a particular emphasis on the growing role of Canadian-owned mining companies in the
region. Chapter 7 by Mexican political economist María Teresa Gutiérrez Haces provides a detailed examination of Canada's relations with
Mexico, the most significant bilateral economic partner Canada has in
the region. Gutiérrez Haces' chapter provides extensive analysis of the
economic and political relations between the two NAFTA partners. She
argues that while much of the literature on the North American region
portrays the two "semi-peripheral" countries as passive victims of US
hegemony, it is important to understand the ways in which Mexico and
Canada have been forging their own relationship, "in which they are
capable of dissenting with, constraining, challenging and leveraging
their neighbor's decisions." The chapter also discusses the particularly fraught period of the NAFTA renegotiations and shows how the
calculations of each partner shifted over time. While there were rocky
moments in the relationship during the renegotiations as both countries faced hostile moves by the Trump administration designed to
divide and conquer them, ultimately the bilateral relationship between
Canada and Mexico was "energized" during this period, argues
Gutiérrez Haces.

Canadian mining investments in Latin America represent the single
strongest economic linkage with the region and the most controversial.
Moore's Chapter 8 makes the case that the Canadian government since
Harper (if not before), and continuing with Trudeau administration,

have followed a policy oriented to serve global commodity markets and maximize mining companies' pursuit of extraordinary profits. At the same time, this policy neglects the social and environmental impacts of these activities for affected communities and, disproportionately, Indigenous peoples in Latin America. The author argues that Canadian diplomacy follows a multilevel approach in support of Canadian mining companies by orienting development cooperation efforts to have domestic mining and environmental legal codes implemented in ways favourable to foreign investors, sign investment protection agreements or include chapters for that purpose into new free trade agreements and allocate advocacy efforts from Canadian embassies throughout the region to support specific mining companies. To make her case, Moore provides examples from actions of Canadian diplomacy at these different levels of influence in Honduras, Mexico, Ecuador, and Colombia in the last two decades, the time when Canadian firms have expanded most rapidly during the commodities boom through the whole of Latin America.

In contrast to that analysis, Haslam's chapter discusses Canada's debate on whether to promote voluntary corporate standards or enforce legal norms on corporate behaviour abroad. The author begins by covering the different claims made by advocacy literature and academics linking Canadian firms' responsibility for human rights abuses with lax Canadian government enforcement of responsible corporate behaviour abroad. Then he reviews what actions the Canadian government has taken to promote voluntary principles of respect for human rights in general terms and corporate social responsibility (CSR) international standards more specifically. Since 2009 until 2020, three different strategies have been implemented or announced, with increasing levels of pressure on Canadian mining firms to maintain good behaviour abroad. None of these seem to be working well, and not necessarily because of that Canadian courts have increasingly taken cases filed by foreigners against Canadian mining firms for human rights violations. Haslam argues that both strategies, one by design, support for CSR activities and investigations of conflicts, and the other by default, the judicial assessment of some cases of human rights violations, comprise a rather weak but expected Canadian state response. Some of the conflicts that are distributional in nature might be easier dealt by pushing for better CSR policies, while other cases of grave human rights violations might be sometimes adjudicated by Canadian courts, then providing compensation and shaming the firm into better behaviour. The sum of both solutions, however, reflects the incomplete Canadian response to problems that, Haslam argues, are often rooted deeper in Latin American societies than the mining projects that apparently bring them to the fore.

Pablo Heidrich and Laura Macdonald

REFERENCE

Cameron, John. 2007. "CIDA in the Americas: New Directions and Warning Signs for Canadian Development Policy." *Canadian Journal of Development Studies* 28, no. 2: 229–49.

Acknowledgments

This edited collection aims to clarify and explain Canada's role in the hemisphere, from diverse theoretical perspectives. The collection provides some explanations for some of the historical weaknesses in the relations between the northernmost country of the Americas and its Latin American partners as well as the opportunities for more egalitarian and mutually beneficial policies. The volume is the product of a five-year Insight grant funded by the Social Sciences and Humanities Research Council of Canada (SSHRC), involving Macdonald, Heidrich, and Christina Gabriel at Carleton University, as well as collaborators in Latin America: María Teresa Gutiérrez Haces in Mexico, Jacobo Vargas Foronda in Guatemala, Alan Fairlie in Peru, and Marcelo Saguier in Argentina. We thank all these researchers, as well as the contributions of Paola Ortiz Loaiza, who provided invaluable research and logistical support, and Federmán Rodríguez, who greatly improved the editing of the volume. That research project culminated in a workshop held at Carleton University on 27 and 28 March 2017 funded by a SSHRC Connections grant. We thank all the people who attended and participated in that workshop, titled "Canada's Past and Future in the Americas." We particularly thank the presenters in that workshop, including Catherine Legrand, Jean Daudelin, Sean Burges, Stephen Baranyi, Marianne Marchand, Geraldina Polanco, Juliana Peixoto, David Boileau, Isidro Morales, Guillaume Fontaine, Ramiro Albrieu, Viviane Weitzner, Jeffrey Davidson, Irvin Waller, Arturo Alvarado, Mónica Serrano, Michael L. Dougherty, Isabel Altamirano, Rachel Vincent, Rene Urueña Hernandez, James M. Lambert, Jorge Schiavon, Chantal Havard, and Sylvia Cesaratto for their important contributions to the discussions, which deeply shaped our thinking about these issues. We also sincerely thank SSHRC, the International Development Research Centre, the Canadian

International Council National Capital Branch, and Carleton University for their financial support. Laura Macdonald also thanks Edgar Dosman, a pioneer in studying and shaping Canada's role in Latin America, for his support and inspiration for her thinking about these issues.

CANADA'S PAST AND FUTURE
IN LATIN AMERICA

1 Introduction: Canada's Past and Future in the Americas: Beyond the "Americas Strategy"

PABLO HEIDRICH AND LAURA MACDONALD

What is Canada's future in the Americas? Despite the fact that Canada is a middle power – like many of the larger Latin American states, with many common foreign-policy objectives – and has been searching for ways to diversify its international relations and trading partners over the last several decades, concrete sign of improvements in the relationship are scarce. This is not a new situation. Historians and political scientists frequently argue that until recently, ties between Canada and Latin America were weak and intermittent. The Mulroney government's decision to occupy Canada's long-empty seat in the assembly of the Organization of American States in 1990 is often seen as a turning point in Canada's relations with the region. Even after that point, however, the relationship has frequently been described as fluctuating and non-committal, reflecting weak historical precedents, but also the absence of strong economic, social, political and cultural ties. As well, successive Canadian governments lacked a sustained commitment to promoting a deeper relationship with their Latin American counterparts. In part this reflects an instinctive long-standing preference on the part of Canadian governments to prioritize their relationship with the United States, their largest market and traditional ally, and with European counterparts that are viewed as Canadian equals. Governments like that of Paul Martin that wished to signal Canada's commitment to help the poorest and most marginalized parts of the world tended to turn to Africa (Cameron 2007).

Stephen Harper's Conservative Party government (2006–15), did prioritize Latin America, with the development of an "Americas Strategy." This strategy, announced by Harper in Santiago, Chile, in 2007, did lead to higher levels of diplomatic engagement with the region and flows of development assistance and more frequent exchanges between Canadian politicians and bureaucrats and representatives from the region.

However, the lack of additional dedicated resources and a one-sided focus on investment and trade linkages meant that, in practice, not as much was achieved as might have been hoped. The expansion of Canadian mining investment in the region did motivate more sustained interest in the region on the part of the private sector and government, but it also ignited widespread protest and considerable controversy about the socio-economic impact of Canadian mining activities. More recently, the decline of commodity prices has highlighted yet again the dangers of excessive reliance on the export of primary goods, a problem that Canada shares with its Latin American neighbours. As discussed in chapters in this volume by Haslam, Deonandan and Schmuland, and Moore, Canadian mining investments are highly controversial because of their social and environmental consequences, and their development impact is limited.

The relationship of Canada with the region has been overshadowed by the elephant in the hemisphere, the United States. For years, the United States was able to treat Latin America as its backyard, particularly the countries of Central America and the Caribbean, and engaged in regular military and economic intervention in the region to promote its own interests. The Canadian government has long tried to stay out of these conflicts to avoid blowback from its superpower neighbour. Nonetheless, its engagement with Latin America has been closely entangled with its bilateral relationship with the United States.

Starting with Canada's decision to join the OAS in 1990, higher levels of engagement with the region were very much motivated by the country's entry into the Canada–US Free Trade agreement (CUSFTA) in 1989, which made apparent the increased importance of north–south flows into the United States' economy and society. Even though Canada was reluctant to extend that agreement to include Mexico, it eventually chose to sign the North American Free Trade Agreement (NAFTA) when it became clear that the United States and Mexico would sign a bilateral agreement if Canada failed to sign on. Canada's joint membership with Mexico in that agreement, in turn, created possibilities for deeper engagement with both Mexico and other Latin American nations for Ottawa. However, Canada's ties with Mexico have often remained half-hearted and conditioned by each country's relationship with the United States, as discussed by Gutiérrez Haces's chapter in this book. This focus on dual bilateralism, and the failure to develop strong North American institutions, meant the potential to create stronger ties was undercut. Over time, as US hegemony in the Latin American region has declined, new spaces have opened up for greater Canadian involvement. With the simultaneous ascendance of China in the global economy and its growing linkages with the developing world, including Latin America, it has been the Asian rising power that has been most able to take advantage of decreased US interest. The question for future Canadian–Latin American relations then becomes whether Canada would play tandem with a retreating United States or take a line of its own in this part of the world.

The election of a new Canadian Prime Minister in October 2015, Liberal Party leader Justin Trudeau, seemed to offer some challenges but also some new opportunities to take advantage of this geopolitical space. The arrival of the Trudeau government initially signalled a move away from Harper's excessive focus on promoting the economic interest of Canadians over concerns with human rights and development in the countries to the south and a return to a liberal internationalist approach to world affairs that would be welcomed by many Latin American states. Six years later, it is still unclear whether the Trudeau government has actually developed a distinct strategy towards Latin America. In fact, several key themes of the previous government remain strong, such as the importance of Canadian mining investment and the desire to sign more free trade agreements. At the same time, the 2016 election of Donald Trump offered serious challenges for both Canada and Latin America. In that context, Canada could take on greater importance for countries of the Americas seeking a friendlier northern partner, but it requires considerable political will to break out of those historic patterns analysed in this collection. And much of the Trudeau government's attention in its first two mandates was absorbed by the difficult issue of the NAFTA re-negotiation talks, which ultimately resulted in the new United States–Canada–Mexico Agreement. At the same time, Canada has also been absorbed with seeking to develop a distinct brand in terms of an "inclusive approach" to trade with some leading economies elsewhere in the world.

This edited collection aims to clarify and explain Canada's role in the hemisphere, from diverse theoretical perspectives. The collection provides some explanations for some of the historical weaknesses in the relations between the northernmost country of the Americas and its Latin American partners as well as the opportunities for more egalitarian and mutually beneficial policies. This introduction sets the scene for the analyses presented by the book's contributors. We begin with an overview of Canada's role in the Americas, and then provide a summary of the record of the Harper and Trudeau governments. The following section provides a brief survey of diverse theoretical perspectives on these relationships and the conclusion sets briefly what the sum of historical experiences and theoretical approaches possibly point for the near-to-medium future in the relations between Canada and Latin America.

Canada in the Hemisphere: Historical Precedents

Canada is clearly located in the Western hemisphere, and before European occupation, Indigenous peoples developed historical connections through the Americas that united "Turtle Island." These historical linkages were largely erased by the distinct colonial powers that occupied the region, destroyed existing civilizations, and imposed new political and economic systems that marginalized Indigenous peoples throughout the hemisphere. In the area that became the Canadian nation-state, the country's leadership and its white settler population identified strongly with Europe and (sometimes in a conflicted fashion) with the United States, which overtook the British as Canada's strongest ally and trading partner. Canada also helped establish and joined many international organizations led by North Atlantic partners, such as NATO and the Bretton Woods institutions. In contrast, early Canadian foreign policymakers tended to downplay or ignore the Latin American and Caribbean countries, even though the former had established formal independence long before Canada itself became an independent nation. Canada in fact had no foreign-policy autonomy from Britain until after the signing of the Statute of Westminster in 1931, which effectively ended its colonial status in the British empire.

Although official Canadian presence in the Americas was limited until the 1990s, businesses, churches, and non-governmental organizations (NGOs) established important social and economic linkages much earlier. As early as the eighteenth century, Canada's participation in the British colonial economy led to strong commercial relations with the British West Indies. British colonies in North America exchanged fish, lumber, and other primary products for sugar, rum, molasses, and spices from the Caribbean colonies as part of the colonial system of triangular trade relations. This colonial economic system led to the establishment of Canadian corporate dominance in aspects of the Caribbean economy, including finance, transportation, insurance, and utilities (Momsen 1992). Canada also eventually became the destination of significant numbers of Caribbean migrants.

Canada's social and economic role in other parts of the Americas where British colonial ties did not operate was more limited. J.C.M. Ogelsby's 1976 book *Gringos from the Far North* documents, however, how Canadian adventurers and entrepreneurs ventured throughout the hemisphere even before Confederation in 1867 in search of fun and profit. Recent historiographical work has shown the significant role played by Canadian missionaries, particularly from Quebec, in the Latin American region and the influence of that experience on the development of the thoughts and practices of Canadian churches, particularly after the emergence of Latin America's liberation theology (Demers 2014). Canada's presence in the hemisphere was certainly not always benevolent. Recent debates about the role of Canadian mining investment were presaged by early works, such as John Deverell and LAWG (Latin American Working Group)'s book *Falconbridge: Portrait of a Canadian Mining Multinational* (1975), which revealed the harmful practices associated with the operations of this Canadian multinational in the Dominican Republic.

Stefano Tijerina (2011; 2019) has also documented the dominance of commercial motivations in early Canadian government policy towards the region, focusing in particular on the case of Canada–Colombia relations. His research shows how the Canadian government undertook a series of actions to promote Canadian trade interests in the region, including the sale of military equipment, and the market-driven logic that dominated Canadian official development assistance (ODA) policies beginning in the 1970s.

In the postwar era, Cold War realities played an important role in shaping Canada's official role in the Western hemisphere. Canada began to deliver development assistance to the British West Indies Federation in 1958, as part of its support for the British-led Colombo Plan in the 1950s, designed to promote economic development and reduce the attraction to the Communist bloc among newly independent states in the Commonwealth (Cermakian 1968, 228). As argued by Jill Campbell-Miller, Canada's early development assistance policy was "firmly embedded in a British Commonwealth and settler colonial mindset that privileged Christian, 'Anglo-Saxon' whiteness over other cultures" (2018, 87). In an early echo of contemporary Canadian aid policy, it also sought to defend capitalist or free-market economic policies with technical assistance, loans, and government investment guarantees.

In contrast with the close ties that were established with the English-speaking Caribbean, Canadian governments felt they had little to gain from involvement in Latin America, an area that the expanding US superpower viewed as part of its back yard. Canada's reluctance to join the Organization of American States (OAS) (and its predecessor, the Pan American Union) contrasted with the country's enthusiasm for membership in innumerable other international organizations as part of its liberal internationalist leanings. For many years, a chair, engraved with Canada's name, gathered dust in the basement of the

OAS. As Peter McKenna argues, Canada's usually dominant foreign-policy style of liberal internationalism and "middlepowermanship" was not applied in the inter-American system, first because of US opposition to Canada's involvement because of its membership in the British Empire. Later, after World War II, Canadian governments decided to stay out of inter-American institutions for various reasons, "including the lack of official government interest in the region, a concern about possible negative implications for the Canada-U.S. relationship, a preference for cultivating Canada's European or North Atlantic connection, and a belief that the UN was more institutionally palatable to Canada" (McKenna 1995, 3).

Canada's desire to avoid hemispheric entanglements was tested, however, by one of the most difficult moments in the hemisphere's (indeed the world's) history. In the early 1960s, hostilities heightened between the US government and the regime of Fidel Castro, which came to power in a revolution in 1959, overthrowing former US ally Fulgencio Batista. In 1961, the US government decided to break diplomatic relations with Havana and impose an embargo on almost all imports of Cuban products as a result of the radicalization of the Castro regime and its nationalization of US properties, as well as its increasingly close ties to the USSR. Progressive Conservative Prime Minister John Diefenbaker, a strong Canadian nationalist, chose to publicly resist US pressure to break relations with Cuba and to endorse the embargo. By maintaining political and economic relations with the Cuban government, successive Canadian governments fuelled Canadians' self-perceptions as maintaining foreign-policy independence fitting of a middle power. For the most part, however, Canada chose to steer clear of involvement in the area, partly out of fear that it would anger the United States, which had established strong political, economic, and military dominance in the region.

One period in which the Canadian government did attempt to grow closer to Latin American countries was during Prime Minister Pierre Trudeau governments. Trudeau's interest in the region was influenced in part by his Québecois heritage and the influence of the concept of Latinité in French Canada, based on the assumption that Quebec shared strong cultural ties with the countries of Latin America (because of their strong Catholic influences and Latin languages). Trudeau had travelled extensively in the region and spoke Spanish fluently (McKenna 1995, 86) and had published op eds and articles about Latin America in the influential journal he co-founded, Cité Libre (Morrison, 2006, 79). He also was looking for potential new markets for Canadian goods, as part of his government's unsuccessful Third Option policy designed to diversify trade relations away from the United States. Trudeau's administration strongly supported the expansion of Canadian state enterprises (crown corporations) via investments of Cameco (uranium mining and nuclear technology licencing), the Canadian Commercial Corporation (arms sales and infrastructure projects), Petrocanada (oil exploration), and supported previous mining investments of Canadian firms, such as INCO (Guatemala), and Falconbridge (Dominican Republic) (Rochlin 1994).

Pierre Trudeau was personally sympathetic to the Cuban regime and friendly with its controversial leader, Fidel Castro. Under Trudeau, Canada joined the Inter-American Development Bank (IDB) and became a permanent observer to the OAS in 1972 but chose not to pursue full membership in the latter organization. The question of Canadian membership then languished for over 15 years, reflecting the lack of strong Canadian public interest or indeed lack of interest on the part of the Latin American states for Canada to join (McKenna 1995, 85–101). The Trudeau government also supported stronger bilateral ties with "hemispheric middle powers" like Brazil, Mexico, and Venezuela. However, perhaps because of the lack of an institutional home for increased contacts with Latin American states, but also because of the continued hegemony of the United States, Canada's political role did not significantly increase, despite growing economic ties. During the 1970s, Canada did not openly oppose US policy in the region, such as its support for the military coup in Chile in 1973 against the democratically elected government of Salvador Allende (Stevenson 2000, 8). Canada's absence from the hemispheric organization represented a major anomaly – the only other Western hemispheric country that was not a member was Cuba, which was excluded after the revolution because of US opposition to its Communist system.

The basis for a shift in the relationship between Canada and the Americas was laid in the 1980s, however. One change was the emergence of strong civil society interest as a result of the emergence of revolutionary movements in several Central American countries, and the Reagan government's militaristic response, which resulted in the escalation of human rights abuses committed by US allies. The Canadian government initially reacted in a cautious manner, reluctant to directly criticize Reagan's policies in the region, but as a result of extensive pressure from Canadian civil society and closer relations with regional middle powers like Venezuela, gradually adopted a more critical posture towards US policy (Baranyi and Foster 2012). Unlike the government of US President Ronald Reagan, which viewed the Central American insurgencies through a Cold War lens, Canadian governments saw them as rooted primarily in social and economic causes. Going against US foreign policy, Canada increased its development assistance to Nicaragua during the left-wing Sandinista government while the United States was funding counterinsurgency forces to overthrow that government (Stevenson 2000, 8). Canada also was the destination of a large number of refugees from the region, fleeing from the dictatorial US allies of El Salvador and Guatemala. Nevertheless, as Brian Stevenson notes, Canadian opposition to US policy in Central America was not particularly vocal, Canada did not take on a leadership role in promoting a resolution of the conflict, it continued to support the governments of US ally regimes in Guatemala, El Salvador, and Honduras, despite their anti-democratic character, and for the most part its policy was "most often quiet, sometimes hesitant, and always overly cautious" (2000, 9).

An important source of convergence between Canada and Latin America during the 1980s was both sides' embrace of the neoliberal economic policies of the so-called Washington consensus (Rochlin 2012, 6). Up to this point, both Canada and major Latin American states had resisted laissez-faire policies and, influenced by Keynesian and structuralist theories, supported the role of state intervention in the economy (although these policies took a different form in Latin America than in Canada, given their distinct social and political realities and forms of insertion into the international economy). However, the recession of the early 1980s led to a rapid shift in economic orthodoxy, promoted by the United States, US-based universities and think tanks, and Washington-based international financial institutions like the International Monetary Fund and World Bank. In Canada, this shift in economic policy, as well as the global economic environment, influenced the decision to sign a free trade agreement with the United States in 1987. Mexico followed Canada's example in the early 1990s, resulting in Canada and Mexico's joint membership in the North American Free Trade Agreement (NAFTA). There was also a degree of political convergence in this period. At the same time that most of Latin America was adopting neoliberal economic policies, the dictatorships that had prevailed in most countries of the region fell, and formally democratic political regimes came to power (still with the major exception of Cuba).

These shifts in the economic and political models in Latin American also led to a rethinking of Canada's traditional lukewarm relationship with Latin America and its institutions. In 1989, the Progressive Conservative Prime Minister Brian Mulroney publicly announced, without earlier consultation or public discussion, that Canada would join the OAS. Foreign minister Joe Clark stated at the time of the decision, "For too long, Canadians have seen this hemisphere as our house; it is now time to make it our home" (McKenna 1995, 136). Clark also began using language about Canada joining the "family" of the Americas that would be used later by Jean Chrétien. The Canadian government indicated that once Canada was a member of the hemispheric organization it could act as a force to promote liberal values of democracy and economic openness in the region, as well as to promote its own narrower interests in trade and investment promotion. After it became a member, Canada took an active role in supporting the consolidation of the organization, for example, by pushing for regular summits, perhaps to compensate for the fact that, unlike the United States, it had weak links with most of the region. Ottawa also supported the creation of the Unit for the Promotion of Democracy (UPD) in 1991, which promoted the adoption and consolidation of liberal democratic electoral practices. Shortly after, the Fujimori *autogolpe* threatened to undermine these fragile democratic practices, and Canadian foreign minister Lloyd Axworthy worked closely with the OAS secretary general Cesar Gaviria to restore democracy after Fujimori resigned and fled to Japan (Cameron and Tockman 2012, 91). Subsequently, Canada and Peru jointly led the effort to create an Inter-American Democratic Charter.

The decision to join the OAS was a result not just of a change of politicians' perspectives but also of the gradual shift within Canada's foreign affairs bureaucracy to take greater interest in the countries of the Americas. Prior to the decision, the Department of External Affairs (as Global Affairs Canada was known in those days) undertook an extensive process of analysis and consultation to develop a comprehensive policy framework which, according to McKenna (1995, 137), was known as "Canada's long-term political strategy for Latin America" (terminology similar to that used under Harper). The strategy included such measures as hosting an annual meeting of the IDB, intensifying bilateral ties with Mexico, participation in the Central American peace process, an increase in bilateral development assistance to Central American countries and the mounting of trade missions to identify new export markets (McKenna 1995, 138). The decision also reflected the belief that the role of the United States had declined in hemispheric affairs and specifically in the OAS, particularly with the end of the Cold War, opening up space for Canada to play a more independent role (a theme which has also characterized more recent Canadian policy as US hegemony has declined even further from its apogee in the early and mid-twentieth century).

Even after the election of Liberal Prime Minister Jean Chrétien in November 1993, many of the same themes were maintained in the new government's foreign policies. The Chrétien government ratified

NAFTA, the agreement with the United States and Mexico negotiated under Mulroney, despite Chrétien's pledge during the campaign to renegotiate the deal. Subsequently, his government signed a free-trade agreement (FTA) with Chile and strongly promoted the Free Trade Area of the Americas (FTAA) initiative launched by Bill Clinton at the Miami Summit of the Americas in 1994. The Summit of Americas held in Quebec City in 2001 was the site of some of the biggest and most vocal protests against the deal, which had become a dead letter by 2004. In this period, Canadian civil society organizations which had begun to collaborate with NGOs and trade unions in the United States and Mexico in a common fight against the NAFTA agreement, began to extend their links with civil society organizations in Latin American countries further south (see von Bülow 2010). The Canadian state also participated in various UN peacekeeping missions in Central America and later in Haiti over the course of the 1990s, reflecting the government's commitment to liberal internationalism, which now was seen relevant in the context of the Americas. The defeat of the FTAA initiative marked a period of decline in Canada–Latin America relations, partly because of the election of a series of post-neoliberal, left-of-centre governments in Latin America that rejected the Washington consensus policies and US leadership upon which the FTAA was based. It was not until the election of Stephen Harper in 2006 that Canada would undertake a new turn towards prioritizing the Americas in its foreign policy.

Harper and the Americas Strategy

Stephen Harper's Conservative Party government came to power in 2006 with minority government status and achieved a majority in the 2010 elections, ultimately losing to Justin Trudeau's Liberals in 2015. With regard to foreign policy, Harper's government was defined by its strong opposition to the liberal internationalism that had characterized Canadian foreign policy behaviour for decades (under both Liberal and Progressive Conservative governments). Harper, however, strongly associated liberal internationalism and middlepowermanship with the Liberal Party brand, and advocated an alternative reading of Canadian history, highlighting themes of "moral steadfastness and martial valour," rather than acting as a helper fixer. He also deeply mistrusted government officials in the Department of Foreign Affairs and International Trade, viewing them as captured by Liberal party ideology. Harper's Conservatives wanted to convert Canada's role in the world to that of a "valiant fighter" (Paris 2014, 275). This commitment to resolute war-readiness had little opportunity for expression in the nations of the Americas, where there has been little military confrontation, but was coupled with a strong commitment to neoliberal economic principles and support for free trade that led to a view of the Americas as an appropriate site for Canadian engagement. Harper's election thus meant that the Canadian state's attention returned to the Americas in a more public way than in earlier iterations but with new assumptions and predispositions.

In his announcement of his government's new Americas Strategy in Chile in July 2007, Stephen Harper lay out three objectives:

- First to strengthen and promote our foundational values of freedom, democracy, human rights and the rule of law;
- Second, to build strong, sustainable economies through increased trade and investment linkages, as well as a mutual commitment to expanding opportunity to all citizens; and
- Third, to meet new security challenges as well as natural disasters and health pandemics[1] (Harper 2007).

Harper's speech criticized the idea that the only choices for peoples of the Americas were "to return to the syndrome of economic nationalism, political authoritarianism and class warfare, or to become 'just like the United States.' This is, of course, utter nonsense," Harper stated: "Canada's very existence demonstrates that the choice is a false one." Harper claimed, further: "It is not in our past, nor within our power, to conquer or to dominate. What we can do is bring forward our resources and goodwill, in concert with those with whom we have historical links as well as significant interests, to build a more democratic, prosperous, and secure hemisphere for all of its citizens. With a new model of partnership at the heart of Canada's approach to the Americas, we know we can strengthen hemispheric cooperation in support of peace, security and development" (Harper 2007).

This speech echoed some traditional Canadian foreign-policy tropes about the country's lack of a colonial past. The ambitious objectives it laid out were supported by action or resources that would justify the claim made in the speech that "Canada is committed to playing a bigger role in the Americas and to doing so for the long term" (Harper 2007). His speech also presented Canada as a possible alternative model to the "socialism of the 21st century" of the Chávez regime in Venezuela, or the extreme free-market position of the United States. In practice, however, Harper's approach was much closer to the US model in promoting greater free trade and private investment, with little attention to the economic, social, and environmental costs of these polices.

Taking on a more prominent role in the Americas was perhaps appealing to the Harper government because it allowed his administration to gain credit in Washington for taking on some responsibility in the region while the superpower's attentions were largely directed elsewhere. A Wikileaks cable revealed that the Harper government was influenced by the Australian government of John Howard's success in "exerting outsize influence with the U.S. in particular – and other powers as well – by emphasizing its relations in its own neighbourhood" (Wikileaks 2009). In this respect, as with the decision to join the OAS, Canada's road to the Americas was an extension of the continentalist logic adopted under the Mulroney administration. However, in many respects, the government failed to live up to the ambitious objectives laid out in Harper's speech.

The following sections examine the Harper government's policy towards Latin America in two of three of the initial pillars of the Americas Strategy: economic policy and support for democracy and human rights. The third pillar, security, is analysed in the chapter by Federmán Rodriguez in this volume.

Economic Relations

It is commonly perceived that the major concrete outcomes of the Americas Strategy were in the area of trade policy. For example, Peter McKenna argues that the centrepiece of the Americas Strategy as of 2012 "appears to be the signing of free trade deals and investment pacts with a handful of countries, with some still in the works" (2012, 359). Canada has signed more FTAs in the Americas than with any other region of the world (seven out of twelve). Agreements have been ratified with Chile (1997), Costa Rica (2002), Peru (2009), Colombia (2012), Honduras (2013), and Panama (2013). More recently, the Harper administration explored membership in the Pacific Alliance but chose not to pursue this option.[2]

The Harper government also placed heavy emphasis on promoting investment by Canadian mining companies. During the commodity boom of 2003–13, Canadian mining companies (particularly exploration firms) expanded rapidly, particularly in the Americas. This rapid expansion gave rise to many conflicts with local communities and was often associated with the violation of the economic and environmental rights of these communities (see the chapters in this volume by Haslam and Deonandan and Schmuland). These conflicts undermined the good image Canada had assiduously cultivated in earlier years as part of its middle-power diplomacy while providing substantial economic gains for Canadian mining firms and their investors, including the Canadian government pension funds and large Canadian private banks (Studnicki-Gizbert 2016).

Canada's trade flows with Latin America grew very rapidly beginning in the 1990s, especially with Mexico, thanks to NAFTA. Trade also grew with Brazil and the Andean nations due to mining investments, with Canada exporting machinery there for those projects and importing their output for refining and global resale. Trade flows have, however, remained imbalanced in favour of most Latin American countries, as the Canadian economy lost competitiveness during the 2000s due to currency appreciation and its own specialization in oil and mineral exports (ECLAC 2007). While most Canadian foreign investment (FDI) in the region was in manufacturing and private–public utilities during the 1990s, reflecting opportunities opened by neoliberal reforms there, the majority of Canadian FDI currently focuses on banking and mining, mirroring the current strengths of Canada's economy (Arellano 2010).

That context explains, together with Harper's vigorously ideological approach to international relations with the Global South, Canada's emphasis on signing free-trade agreements with investment protection clauses and a generally conservative attitude to dealing with policy changes in Latin America. After all, the region has become increasingly relevant for Canadian firms' profits and their investors, and those gains are not related to the social or economic progress of the region, but with Canada's role in facilitating the largest possible profits from mineral or financial extraction (Heidrich 2016). That fact has not escaped the attention of the Trudeau government since coming to office in 2015 since it has continued, by and large, the same type of economic diplomacy of the Harper years, wrapped in a more progressive language.

Multilateralism, Democracy, and Human Rights

As discussed above, Canada came late to the OAS party, and while the Harper government has maintained its financial support for the organization (Canada has become its second-largest funder), neither the OAS nor the promotion of human rights and democracy was a major area of focus, in contrast with Canada's behaviour when it first entered the organization. Part of this weakness is a result of the decline of the relevance and effectiveness of the OAS, as Latin American and Caribbean states began to create new regional organizations, like the Union of South American Nations (UNASUR), created in 2008, and the Community of Latin American and Caribbean Nations (CELAC), created in 2011. These organizations were explicitly founded as alternatives to the role of the OAS in regional governance, and consciously exclude the United States and Canada. The OAS was portrayed by post-neo-liberal Latin American regimes as a relic of the era of US dominance, and Canada was linked with the United States in the minds of leftist Latin American leaders. One Foreign Affairs official was quoted in a WikiLeaks document saying that, "The Latinos don't want the US at the table, and they see Canada as an extension of the US" (Wikileaks 2009; Shane 2011). As argued by Cameron and Tockman (2012, 93), the tight-rope act that Canada has wished to perform in the hemisphere between reconciling its close relationship with the United States with playing the role of a good multilateral citizen became increasingly difficult as more states moved away from the Washington consensus policies of the 1990s.

Although the Harper government is not responsible for shifts in the Latin American region that have resulted from the decline of the role of the United States and the growing search for post-neoliberal regional fora, it has, at points, aggravated the situation by siding with the United States on controversial themes. At the 2012 Summit in Colombia, for example, Canada's participation was described as largely dominated by trade issues. The Harper government's participation in other aspects of the Summit has been described as "thin": Harper participated in a CEO forum within the summit, and a separate CEO forum between Canada and Brazil was held. However, Canadian participation in civil society forums, which earlier governments had worked to promote, was lim-ited, and Harper had few meetings with his counterparts (Ditchburn 2012). More seriously, Harper and US President Barack Obama were seen as isolated from the rest of the leaders of the region around their refusal to invite Cuba to join the organization, against the wishes of the other members. Harper was unapologetic, stating, "I think we have taken a principled position ... And when we have taken principled positions, we are prepared to argue that and discuss them. But obvi-ously we don't have our positions dictated either by any one country, or frankly, by any group of countries." Carlo Dade, former Executive Director of the Canadian Foundation for the Americas (FOCAL), a think tank which lost its government funding under the Harper admin-istration, explained Harper's position: "These guys personally don't like Cuba. They don't like Communists. And so, they're still fighting the Cold War."

President Obama's rapprochement with Cuba forced Harper to change position on this issue, and he reluctantly endorsed Cuban entry in time for the Panama Summit in 2015. And Canada's Department of Foreign Affairs and International Trade (DFAIT) did play an important role in hosting behind-doors meetings with US and Cuban government officials that led to the eventual reopening of relations between the two countries. In addition, the Harper government also focused on such security initiatives in the OAS, such as the Canadian Initiative for Security in Central America, which offers police and border-security training to officials in countries struggling with crime and illegal immigration, reflecting the government's militarism and tough-on-crime domestic policies (CTV News 2012).

Canada has also faced considerable criticism for its position on the June 2009 coup and subsequent events in Honduras. While Canada, like other states of the hemisphere, initially condemned the coup, the Canadian government did not impose any type of sanctions on the regime that replaced the democratically elected government of Manuel Zelaya. And once the United States did an about-face and stated that it would recognize the results of the November 2009 election, even if Zelaya was not returned to power before they took place, Canada then backed the US position. The Harper government recognized Porfirio Lobo as president as a result of those elections, despite the fact that there was no international observation and the fact that they were carried out in a context of widespread human rights violations. Honduras remained a priority aid recipient for Canada in the region after the elections. Cameron and Tockman (2012) argue that the Harper government's position on Honduras was highly ideological, motivated both by its distaste for the Venezuelan-led Bolivarian Alliance for the Americas (ALBA) that was leading opposition to the Honduran government and by the significant Canadian economic interests in the country. Peter Kent, the secretary of state for the Americas, in fact implied that Zelaya had acted in an unconstitutional manner, supporting the argument made by the perpetrators of the coup, even if he continued to condemn the coup itself (Cameron and Tockman 2012, 101–7; see also Shipley 2013).

Overall, the Harper government's approach to human rights and democracy issues in the Americas was highly partisan, showing a tolerance for human rights abuses carried out in ideologically friendly states, while engaging in strong condemnation of the human rights record of other governments, like that of Chávez in Venezuela, to which it was ideologically opposed. The Harper government also was widely criticized by civil society actors and academics in both Canada and Latin America for its support for the activities of mining companies based in Canada, which were perceived as running roughshod over the rights of local (especially Indigenous) residents, as discussed in other chapters in this volume.

Justin Trudeau's Liberal Government – Change or Continuity?[3]

When Liberal leader Justin Trudeau came to power in 2015, he promised that his government would undertake a dramatic policy shift away from Harper-era policies, both in domestic and in international policies. Even if there are strong rhetorical and foreign-policy differences between Harper's and Trudeau's regimes, however, there are also some notable elements of continuity, particularly with regard to their policies in this hemisphere. The Liberal government has boldly proclaimed a return to principles of liberal internationalism, and the former foreign minister, Chrystia Freeland's, main foreign-policy speech in the House of Commons declared Canada's support for principles of multilateralism, democracy, human rights, gender equity, and environmental sustainability, and distanced Canada from the policies of the Trump administration on both climate change and international trade (Global Affairs Canada 2017a). These commitments were undoubtedly welcome to most Latin American states.

However, the Trudeau government did not openly prioritize relations with Latin America (or any other world region), perhaps in part because the election of President Trump and the NAFTA re-negotiations monopolized much of its attention. Beyond this, the return to a liberal internationalist foreign-policy orientation means that the Canadian government is unlikely to prioritize Latin America over other areas of the developing world with greater need for humanitarian and financial support, particularly in Africa and parts of Asia. Trudeau travelled to Latin America to attend the Summit of the Americas in Lima in April 2018 but did not distinguish himself with any prominent announcements and left early to deal with a domestic controversy around oil sands production and the construction of a pipeline (Macdonald and Pickup 2018). Much of the Trudeau government's energy and attention was diverted from broader foreign-policy issues to the NAFTA re-negotiation talks and later the COVID-19 crisis.

Beyond NAFTA, the Trudeau government largely followed his predecessor's commitment to expanding trade and investment in the region, and particularly with the Pacific Alliance group of Chile, Colombia, Mexico, and Peru. Canada signed a Joint Declaration on Partnership with the Pacific Alliance members in 2017 and has been in negotiation with the group to become an associated state. The government tried to restart trade negotiations with the MERCOSUR regional bloc, presently composed of Brazil, Argentina, Uruguay, and Paraguay. As well, it continues to promote Canadian investments in mining, banking, and privatized public services in Latin America, including private health services. In accordance with electoral promises made in 2015, the trade minister announced in January 2018 the creation of an independent Canadian Ombudsperson for Responsible Enterprise (CORE), mandated to independently investigate, report, recommend remedy, and monitor compliance with the government's policies around the responsible behaviour of Canadian corporations abroad (Global Affairs Canada 2018). However, it has not provided any personnel or budget to that office, rendering the fulfilment of this promise moot and effectively continuing the policies of the Harper governments to lightly encourage mining firms to respect human rights abroad.

Another distinctive element of the Liberal government's policy is its commitment to a feminist foreign policy and its announcement of a Feminist International Assistance Policy (FIAP) in June 2017. The FIAP reformed Canada's traditional development assistance policies to focus intensively on promoting gender equality and the rights and empowerment of women and girls. As part of this policy, the government eliminated the practice of identifying a fixed list of countries as priority aid recipients. The policy signalled that the government would increase support for least-developed countries, with a particular focus on sub-Saharan Africa (Global Affairs Canada 2017b). Nevertheless, there is considerable inertia in regional allocations of development assistance, which means there is unlikely to be a dramatic shift away from aid to middle-income countries in Latin America. In 2019, total Canadian development assistance to Latin America amounted to 13.86 per cent of the total development assistance budget, a considerable decline from the high level it reached under Harper in 2012 of 17.3 per cent, when that level was inflated by the government's support for Haiti after the earthquake in 2010 (CIDP 2020). As well, the relative strength of the women's movement in Latin America means that it is likely the region will see benefits from the new Feminist International Assistance Policy. On 4 June 2018, the government announced almost $80 million in development assistance for Latin America, targeted at empowering women and girls (Government of Canada 2018).

Overall, despite some promising elements in the Trudeau government's policies, there has been more continuity than change from the policies undertaken under Harper. This is particularly clear in the one area in which the Trudeau government has shown strong interest in Latin America, which is its response to the collapse of democratic institutions in Venezuela since 2014, under the presidency of Nicolás Maduro. Canada is a founding member of the Lima Group, an informal multilateral grouping formed at a meeting in Lima, Peru, in 2017 after the attempted shutdown of the Venezuelan Congress and the failure of the OAS to approve a text condemning this development.[4] While the Lima Group initially presented itself as a mechanism to drive diplomatic pressure on the Maduro administration, its discourse shifted to be more hardline in 2019, demanding the resignation of Maduro and the installation of opposition leader Juan Guaidó as acting president for a transitional government (Chaves Garcia 2020).

Despite Canada's participation in the Lima Group and its moves beyond that, such as sanctions placed on representatives of the Maduro regime, it has thus far failed to obtain its stated goal of a restoration of electoral democracy in Venezuela. In fact, pressures from Canada, other countries from the Lima Group, and the United States have not slowed the mounting social, economic, and humanitarian disaster faced by most Venezuelans, as Maduro stubbornly clings to power and has increased its levels of repression. Canada's role has been heavily criticized by the Maduro (and formerly Chávez) government, and by some Canadian academics and civil society activists sympathetic with the Maduro government, who see this development as a sign of the increasingly interventionist character of the Canadian government in Latin America (see Kingsbury 2019).

The reasons why Canada has adopted such a prominent role in the otherwise mostly South American Lima Group are unclear. They could relate to the depth of the economic and humanitarian crisis in that country, perceived in Canada as an element that could destabilize other neighbouring countries via violence spillovers or large flows of economic refugees. Domestically, the presence of an articulate and rapidly growing Venezuelan diaspora in Canada has also been pushing the Canadian government to take on a more prominent role in support of the opposition, and even personal relationships between MPs and Global Affairs officials, with prominent members of the Venezuelan opposition, like Leopoldo López, playing a possible role. Finally, former foreign minister Chrystia Freeland's strong ideological commitment to liberal internationalism, could be contributing to what Rodríguez, in his chapter of this volume, calls "sanctimonious idealism." Regardless of the evident strength and sincere motivations of Canada's commitment to a restoration of democracy in Venezuela, the result so far is an undeniable failure of its strategy to displace the Maduro regime. Meanwhile, the ongoing suffering of the Venezuelan people suggests that a change in approach is badly needed, as we discuss in the conclusion of this volume.

Theoretical Approaches

Finally, we provide a brief overview of some of the theoretical lenses that have been applied to the study of Canada–Latin America relations in recent years. This is far from an exhaustive account of the literature: We merely highlight some of the main approaches and identify some representative works within each. There has been little academic work that surveys these alternative perspectives, and much literature on the topic is atheoretical. Nonetheless, we can discern some contending views (for a more extended discussion of the influence of contending theoretical perspectives on Canadian policy, see the chapter by Rodrí-guez in this volume).

A realist approach to international relations has rarely been applied to the study of Canada–Latin America relations (and is rare in the study of Canadian foreign policy more broadly), partly because there are few instances of state-to-state security threats in the region, and because in the post–Cold War era in particular, Canadian security interests are not much affected by events there. Jean Daudelin (2003, 637) presents a realist-inflected perspective that would exemplify this approach: "Beyond the bilateral relationship with the United States, the Americas are essentially marginal to Canadian interests, however one defines the latter, and they are likely to remain so for the foreseeable future. In such a context, the challenge for the government is to design a policy that is consistent over time, that can have significant impact in the few areas that are more relevant to the country's interest, that is likely to command sufficient resources for that purpose and can be sustained over time. No easy task."

Daudelin judges even Canadian economic interests in the Americas beyond the United States to be quite limited based on low levels of trade and investment with that area: "in economic terms, Latin America and the Caribbean can in no way be conceived as important nor even relatively significant region for Canada, not now, not in the future" (2003, 641). A realist perspective's emphasis on the lack of strong Canadian interests in the region could certainly help explain the failure of the Canadian government to engage in a sustained commitment to increasing its role and influence in the region (although Yvon Grenier [2019, 58] adopts the opposite perspective in arguing that Canada's bold position on Venezuela regarding human rights and democracy promotion can partly be explained by the weakness of Canadian interests there). Daudelin provides a refreshing alternative to some of the exaggerated claims made by liberal academics and foreign-policy decision-makers like Chrystia Freeland about Canada's indispensability in promoting peace and democracy and multilateral solutions in the world. The realists' state-centred perspective, however, does little to explain many of the changes over time in Canadian policy. It makes it difficult to explain, for example, why the Canadian state does at times adopt positions distinct from that of the United States, where its real interests would seem to lie, such as when the government failed to follow the US position on Cuba or on Central America in the 1980s. It overlooks the role of specific domestic actors, particularly mining companies, that may achieve disproportionate influence over government decision-making[5] (see Moore in this volume), as well as the role of ideational factors (see, for example, Asa McKercher's chapter in this volume; see also Macdonald 2008), and the role played by civil society actors (see Foster and Deonandan and Schmuland in this volume) or diaspora communities (see Gabriel and Macdonald in this volume).

A more common theoretical approach to understanding Canada's role in the Americas is a liberal internationalist perspective. From this perspective, Canada, as a middle power in the international system, can and does adopt positions distinct from that of the United States in Latin America and elsewhere, in pursuit of its high-minded ideals, such as democracy, human rights, and the construction of a strong rules-based order. An example of such an approach is presented by Brian Stevenson's 2000 book, *Canada, Latin America and the New Internationalism: A Foreign Policy Analysis*. In this book, Stevenson argues in favour of the idea that Canada's approach to the region is framed by the country's middle-power status and internationalist perspective but that concrete policies are an outcome of the interaction between the international system (pressures coming from the United States) and domestic politics (the increasing role of Canadian social movements and non-governmental organizations [NGOs] pushing for a policy based on support for human rights and democracy and a critique of US policy). Stevenson's book only deals with the period up to 1990 and therefore does not provide an adequate guide to more recent policy. He forged new ground in highlighting the significance of NGOs as one of the relevant actors in explaining Canada's policy in the region, but largely overlooks the role of corporate interests, which became more apparent in the Harper era.

Stephen Randall's 2008 evaluation of the state of the relations between Canada and the Americas also highlights societal changes, such as the increased study of Spanish in Canadian schools, the rise in Latin American immigration, and the growth of economic ties. He also argues for the importance of shifts in Canadian public opinion on trade, human rights, and drug policy and predicts a growing importance of the region in the country's political consciousness as a result. A more recent assessment by Peter McKenna (2018) which is in line with a liberal perspective argues that Canada has fallen down in its approach to the Americas precisely because it has failed to learn the lessons of the past about the importance of people-to-people contact, of listening to Latin American's perspectives, and the capacity of Canada to "punch above its weight" in Latin America given its numerous strengths in such areas as oil and gas, mining, tourism, elections monitoring, rights protection, communications technologies and renewable energy. He argues, "Canada just needs to be more demonstrative, to rid itself of its 'follower mentality', and to make a concerted effort to make its presence felt in the region. Again, the critical importance of political will comes into play here" (2018, 28). The contributors to the 2011–12 *Canada Among Nations* (Bugailiskis and Rozental 2012), which focuses on the Canada–Mexico relationship, also largely embrace a liberal perspective on the mutual benefits to both Canada and Mexico that flow from expanded economic and social ties, without examining the power imbalances in the relationship or the negative impact of some forms of Canadian corporate behaviour abroad.

While a liberal approach contributes a useful appreciation of the importance of multilateral institutions, economic interdependence, values, and non-state actors, it tends to overlook the asymmetries in power relations among states, corporations, and non-business civil society actors. It also provides little explanation as to why the Canadian state has not consistently and coherently promoted democracy and human rights in Latin America, under both Conservative and Liberal administrations. A critical theoretical perspective offers more insight into these issues, highlighting the continued role of US intervention, the predominance of corporate interests in many cases of foreign-policy decision-making, and the relative weakness of value-based arguments from civil society. A strong debate has emerged among critical observers of the region and Canada's role within it in recent years, however. In the past, some left nationalist authors (see, for example, Levitt 1970; Laxer 1973; Watkins 1989) viewed Canada as a dependent power, almost inevitably submitting to or complicit in US intervention in Latin America. In recent years, the decline of US hegemony and the rise of Canadian mining investment in the region has led to a new approach. This position argues that Canada must be considered not a dependent power or a relatively independent middle power but as an imperialist (if secondary) power in its own right (see Kellogg 2005, 2015; Gordon and Webber 2008, 2016, 2019; Klassen 2009, 2014; Shipley 2013). These authors argue that the Canadian state has acted in a "coherent, systematic and self-reproducing" fashion to promote the interests of Canadian capital by supporting right-wing regimes in the hemisphere, such as that of Honduras and Colombia; intervened in the affairs of left-wing governments; and supported the destructive behaviour of Canadian extractivist industries and other exploitative companies (Gordon and

Webber 2019, 80). This approach has led to a healthy debate on the motivations and impact of Canadian foreign policy. However, in our view, it exaggerates the relative importance of Canadian capital in Latin America compared to that of other Western states, non-Western powers (i.e., China), domestic elites in those countries, and the coherence of Canadian foreign policy in "resolutely" promoting capitalist interests in the region. In contrast, it tends to downplay the agency of Latin American states, the variety of its domestic political forces, and generally portrays local populations as passive victims of the implacable machinations of the Canadian imperialist regime.[6] Moreover, its economistic leanings means it overlooks non-material dimensions of policy, such as the role of race, ethnicity, and gender in driving the behaviour of white settler countries like Canada, an approach that has been developed further by feminist and post- and de-colonial authors (see, for example, Altamirano-Jiménez 2013).

Conclusion: Beyond the Americas Strategy?

The contributions to this collection first survey in greater detail historical backgrounds for Canada's relations with Latin America and later move on in a thematic fashion to assess the contributions and limitations of the Harper government's Americas Strategy, as well as the subsequent policies of the Trudeau government. In the conclusion, we briefly outline the record of the Justin Trudeau government to date and provide some possible scenarios regarding where Canada will take its engagement with Latin America in the near-to-medium term.

NOTES

1 The third pillar eventually changed to "Fostering lasting relations," which emphasized "strengthening bilateral and multilateral relations throughout the hemisphere across government, the private sector, civil society, academia and communities." Goal 2 was changed to include both addressing insecurity as well as promoting "freedom, democracy, human rights and the rule of law" http://www.international.gc.ca/americas -ameriques/assets/pdfs/strategy-eng.PDF.

2 The Trudeau government signed a "Joint Declaration of a Partnership between Canada and the Members of the Pacific Alliance" in June 2016 and is in talks regarding a free-trade agreement with the region.

3 Part of this section is taken from Macdonald 2018.

4 The membership of the Lima Group is in flux. Argentina, Bolivia, Mexico, and Peru have withdrawn from participation in the group, and as of November 2021, the remaining members are Brazil, Canada, Chile, Colombia, Costa Rica, Guatemala, Guyana, Haiti, Honduras, Panama, and Paraguay.

5 Although he does recognize that policies in fringe areas can be "hijacked" by powerful actors like Bombardier in the relationship with Brazil, for example (2003, 659).

6 For more discussion of this debate, see Garrod and Macdonald 2016 and a rejoinder by Gordon and Webber 2019.

REFERENCES

Altamirano-Jiménez, Isabel. 2013. *Indigenous Encounters with Neoliberalism: Place, Women, and the Environment in Canada and Mexico.* Vancouver: UBC Press.

Arellano, J.M. 2010. "Canadian Foreign Direct Investment in Latin America." Background paper. Prepared for the Dialogue on Canada–Latin American Economic Relations, 27–8 May 2010. The North-South Institute (May).

Baranyi, Stephen, and John Foster. 2012. "Canada and Central America: Citizen Action and International Policy." In *Canada Looks South: In Search of an Americas Policy,* edited by Peter McKenna, 240–64. Toronto: University of Toronto Press.

Bugailiskis, Alex, and Andrés Rozental, eds. 2012. *Canada Among Nations 2011–2012: Canada and Mexico's Unfinished Agenda.* Montreal and Kingston: McGill-Queen's University Press.

Campbell-Miller, Jill. 2018. "Encounter and Apprenticeship: The Colombo Plan and Canadian Aid in India, 1950–1960." In *A Samaritan State Revisited: Historical Perspectives on Canadian Foreign Aid, 1950–2016,* edited by Greg Donaghy and David Webster, 27–52. Calgary: University of Calgary Press.

Canadian International Development Platform (CIDP). 2020. "Canada's Foreign Aid." Accessed 18 October 2021. http://cidpnsi.ca/canadas-foreign-aid-2012-2/.

Cermakian, Jean. 1968. "Canada's Role in the Foreign Aid Programmes to the Developping [*sic*] Nations: A Geographical Appraisal. *Cahiers de géographie du Québec* 12, no. 26: 225–34.

Chaves Garcia, Carlos Alberto. 2020. "La crisis política en Venezuela y el papel del Grupo de Lima: balance y desafíos de su acción diplomática." *Revista de Relaciones Internacionales, Estrategia y Seguridad* 15, no. 1: 177–93.

Daudelin, Jean. 2003. "Foreign Policy at the Fringe: Canada and Latin America." *International Journal* 58, no. 4: 637–66.

Demers, Maurice. 2014. *Connected Struggles: Catholics, Nationalists, and Transnational Relations between Mexico and Quebec, 1917–1945*. Montreal/ Kingston: McGill-Queen's University Press.

Deverell, John, and LAWG (Latin America Working Group), 1975. *Falconbridge: Portrait of a Canadian Mining Multinational*. Toronto: James Lorimer & Co.

ECLAC. 2007. "Chapter 4: Canadian FDI in Latin America." In *Foreign Investment in Latin America*, 139–90. Santiago, Chile.

Garrod, J.Z., and Laura Macdonald. 2016. "Imperialism or Something Else? Rethinking 'Canadian Mining Imperialism' in Latin America." In *Mining in Latin America: Critical Approaches to the "New Extraction,"* edited by Kalowatie Deonandan and Michael Dougherty, 100–15. Milton Park, Abingdon: Routledge.

Global Affairs Canada. 2017a. "Address by Minister Freeland on Canada's Foreign Policy Priorities." (6 June 2017). Accessed 18 October 2021. https://www.canada.ca/en/global-affairs/news/2017/06/address_by _ministerfreelandoncanadasforeignpolicypriorities.html.

Global Affairs Canada. 2017b. "Canada's Feminist International Assistance Policy." Accessed 18 October 2021. http://international.gc.ca/world -monde/issues_development-enjeux_developpement/priorities-priorites /policy-politique.aspx?lang=eng.

Global Affairs Canada. 2018. "The Government of Canada Brings Leadership to Responsible Business Conduct Abroad." Accessed 18 October 2021. https://www.canada.ca/en/global-affairs/news/2018/01/the _government_ofcanadabringsleadershiptoresponsiblebusinesscond.html.

Gordon, Todd, and Jeffery R. Webber. 2008. "Imperialism and Resistance: Canadian Mining Companies in Latin America." *Third World Quarterly* 29, no. 1: 63–87.

– 2016. *Blood of Extraction: Canadian Imperialism in Latin America*. Black Point, NS: Fernwood Publishing.

– 2019. "Canadian Capital and Secondary Imperialism in Latin America." *Canadian Foreign Policy Journal* 25, no. 1: 72–89.

Government of Canada. 2018. "Canada Announces $79.21 Million in Development Assistance for Americas." Accessed 19 October 2021. https:// www.canada.ca/en/global-affairs/news/2018/06/canada-announces-7921 -million-in-development-assistance-for-americas.html.

Heidrich, P. 2016. "Determinants, Boundaries, and Patterns of Canadian Mining Investments in Latin America (1995–2015)." *Latin American Policy* 7, no. 2: 195–214.

Jayanti, Subbarao Venkata. 1991. "The Impact of Latin American Debt Crisis on U.S., U.K., and Canadian Bank Stocks." LSU Historical Dissertations and Theses.

Kellogg, Paul. 2005. "Kari Levitt and the Long Detour of Canadian Political Economy." *Studies in Political Economy* 76: 31–60.

– 2015. *Escape from the Staple Trap: Canadian Political Economy after Left Nationalism.* Toronto: University of Toronto Press.

Kingsbury, Donald. 2019. "From Middle Power to Regime Change Specialist: Canada and the Venezuela Crisis." *NACLA Reporting on the Americas.* Accessed 19 October 2021. https://nacla.org/news/2019/05/02/middle -power-regime-change-specialist-canada-and-venezuela-crisis.

Klassen, Jerome. 2009. "Canada and the New Imperialism: The Economics of a Secondary Power." *Studies in Political Economy* 83: 163–90.

– 2014. *Joining Empire: The Political Economy of the New Canadian Foreign Policy.* Toronto: University of Toronto Press.

Laxer, Robert. 1973. *(Canada) Ltd.: The Political Economy of Dependency.* Toronto: McClelland and Stewart Limited.

Levitt, Kari. 1970. *Silent Surrender.* New York: St. Martin's Press.

Macdonald, Laura. 2018. "La política exterior de Canadá hacia América Latina, de Harper a Trudeau: ¿un regreso al internacionalismo de potencia intermedia? *Revista Mexicana de Política Exterior* no. 114 (September–December): 177–98.

Macdonald, Laura, and Megan Pickup. 2018. "A Missed Opportunity: Canada's Shallow Participation in the Summit of the Americas – and the Americas." McLeod Group. Accessed 19 October 2021. http://www .mcleodgroup.ca/2018/04/19/a-missed-opportunity/.

McKenna, Peter. 1995. *Canada and the OAS: From Dilettante to Full Partner.* Ottawa: Carleton University Press.

– 2018. "Canada and Latin America: 150 Years Later." *Canadian Foreign Policy Journal* 24, no. 1: 18–38.

Momsen, Janet Hensall. 1992. "Canada-Caribbean Relations: Wherein the Special Relationship?" *Political Geography* 11, no. 5: 501–13.

Morrison, David. 2006. *Aid and Ebb Tide: A History of CIDA and Canadian Development Assistance.* Waterloo, ON: Wilfrid Laurier University Press.

Rochlin, James. 1984. *Discovering the Americas: The Evolution of Canadian Foreign Policy Towards Latin America.* Vancouver: UBC Press.

Shipley, Tyler. 2013. "The New Canadian Imperialism and the Military Coup in Honduras." *Latin American Perspectives* 40, no. 5: 44–61.

Stevenson, Brian J.R., 2000. *Canada, Latin America, and the New Internationalism: A Foreign Policy Analysis, 1968–1990.* Montreal/Kingston: McGill-Queen's University Press.

Studnicki-Gizbert, Daviken. 2016. "Canadian Mining in Latin America (1990 to Present): A Provisional History." *Canadian Journal of Latin American and Caribbean Studies/Revue canadienne des études latino-américaines et caraïbes* 41, no. 1: 95–113.

Tijerina, Stefano. 2019. "One Size Fit All? Canadian Development Assistance to Colombia, 1953–1972." In *A Samaritan State Revisited: Historical Perspectives on Canadian Foreign Aid*, edited by Greg Donaghy and David Webster, 123–43. Calgary: University of Calgary Press.
– 2012. "One Cinderblock at a Time: Historiography of Canadian-Latin American and Colombian-Canadian Relations." *Desafíos* 24, no. 1: 275–92.
von Bülow, Marisa. 2010. *Building Transnational Networks: Civil Society and the Politics of Trade in the Americas*. Cambridge: Cambridge University Press.
Watkins, Melville. 1979. "Staples Redux." *Studies in Political Economy* 79 (1989): 16–35.

2 Locating Latin America: Geography, Identity, and the Americas in Canadian Foreign Policy

ASA McKERCHER

Writing in 1947, Vincent Massey, until recently the Canadian representative in London, reaffirmed his longstanding opposition to Canada's involvement with the Pan-American Union (PAU). The lack of Canadian membership in this hemispheric body was considered a sign of Canada's disinterest in Latin America, or what Massey termed a "friendly detachment." As the future governor general explained, such detachment sprang from the fact that, for Canadians, "The western hemisphere will always mean less to us than the northern hemisphere." The following year, Massey estimated that the Western hemisphere was not "a unit in a geographic sense," for Canada was physically separate from the rest of the Americas, with the obvious exception of the United States, a country that dominated much Canadian thinking about foreign affairs. In addition to this geographic gulf, he judged, too, that there were "striking cultural differences between the nations of North and South America." Such differences did little to foster a common hemispheric identity. In his overall opinion, since Canada had "much more in common both culturally and politically" with European states and the "sister nations" of the British Commonwealth than with Central and South America, it belonged "to the northern hemisphere rather than to the western, for in the northern half of the globe are both Great Britain and the United States" (Massey 1947; 1948, 694). While clearly an over-generalization, Massey's sense of Canada's geographic position and cultural orientation reflected a view dominant among the country's foreign policymaking elite throughout much of the twentieth century. The stress on Canada as a trans-Atlantic country is a notion deeply rooted not only in Canadian geography, but also in cultural identity. And as Massey had readily admitted, locating Canada and Canadians outside of the Western hemisphere and apart from Latin Americans helps to explain the country's historical – and even continued – aloofness from the region.

Mental Maps and Canada's Geopolitical Position

The way in which countries are situated within the world is important. Positioning Canada as a northern nation, for instance, implies a connection to events in the Arctic. In turn, attendant mythologizing about Canadians as a northern people helps to account for Canadian concerns with protecting Arctic sovereignty. "English-speaking Canadians," geographer Cole Harris observed a half-century ago, "tend to explain themselves in terms of land and location," with the North an obvious and important point of reference (Harris 1966, 27; Berger 2010, 913–30; Cavell 2002, 364–89). As the northern example indicates, geographic framings involve more than a focus on territory, with land and identity frequently mixed together. Thus, this process of situating a country is not only reflective of geography but of more mutable cultural factors. To take another example, one might consider what is implied culturally, politically, and economically by viewing Canada as a Western country. "An uncertain national identity has been, and remains, a major theme in Canadian history," notes historian Robert Bothwell, "and has been reflected in Canada's foreign relations" (Bothwell 2007, 4–9). Historically, many Canadians, like Massey above, were Anglocentric in outlook and saw their country as a part of a North Atlantic Triangle along with Britain and the United States, two countries that loom large in Canadian culture, economics, politics, and foreign policy (Brebner 1945; Granatstein 1985; Nossal 1992; Vucetic 2011). Additionally, Canada was once a part of the British World, an imagined community centred around British historical and cultural identities that remained strong among Canadians until the 1960s (Buckner and Francis 2006). Benedict Anderson's concept of nations as imagined communities is important here because the way Canada is framed as a Western, northern, Atlantic, or English-speaking nation not only underscores what Canada is but also what it is not (Anderson 1991, 5–7). Policymakers' sense of place has an important influence on their actions, reflecting how they see the world and Canada's place internationally. In terms of Latin America, this paper makes clear – as did Massey – that geographic and cultural location have been key elements of the way in which many Canadians have framed their involvement with the region and helps to explain why foreign policymakers have tended to ignore the Western hemisphere.

In exploring the placement of Canada in relation to Latin America, below I utilize the concept of "mental maps," that is, the ways in which foreign policymakers situate themselves and their state within the world, which in turn affects how they view international relations.

Influencing policymaking, this "mental geography" mixes spatial and cultural understandings to create "a composite – of past experience, present observation, and future expectation" about where a state should involve itself and where its interests lie (Henrikson 1980, 496–505; Casey and Wright 2008; Scott 2011; Barney 2015). Mental maps, in short, reflect understandings of geopolitics rooted in location and identity. In his own work on mental maps and Canadian foreign policy towards the "Third World" – a Cold War neologism laden with geographic and cultural meaning – David Webster has emphasized that policymakers viewed this vast and varied region as "peripheral, and kept their gaze locked tightly on the North Atlantic core." Thus, it was left up to Canadian non-state actors to take a more prominent role in the post-colonial world (Webster 2016, 156).

With Latin America, Webster's observations ring true: Foreign policymaking elites and commentators on foreign affairs generally saw the Western hemisphere as *terra incognita* and as marginal to Canada's interests, with non-state actors – many of them francophones, who took a different geopolitical view – stepping into the breach (see Braz 2016, 349–61). What one finds, then, is that the recent observation of two political scientists that Canadians, at present, "appear to be relatively indifferent toward their hemispheric identity," is a notion firmly rooted not only in Canada's history but in mental geography (Thérien and Mace 2013, 159). As Massey's comments above illustrate, Canadian officials framed the issue in this way, and while spatial concepts did not in themselves determine foreign policymaking towards Latin America, they certainly contributed to the inattention accorded to the region by officials in Ottawa. Below, I offer an overview of the history of Canadian relations with Latin America, albeit through a geospatial lens.[1] Overall, examining Latin America's place in Canadian mental maps serves to trace the downs and ups of Canadian interest in the region that is at the core of this volume. While Stephen Harper's Americas Strategy marked an important rhetorical commitment to the Western hemisphere, the shallowness of his government's resulting policy should not be surprising. After all, the Conservatives cast themselves as defenders of a "lost" Canadian identity tied to Canada's British past (Taber 2011), an identity shared with Massey and which continues to colour Canadian mental maps.

Early Encounters with Latin America

For much of Canada's early history, there was little official interest in Latin America. Rather, those Canadians who sought ties to the Western

Hemisphere were concerned with making money or saving souls: Businessmen and members of religious orders were the main Canadian actors in the region, and both groups had a significant impact within their respective spheres. Meanwhile, the smattering of Canadian trade commissioners – six in the region prior to the Second World War – gives an indication of government priorities (Armstrong and Nelles 1988). This situation was unsurprising, perhaps, as Canada lacked control over its foreign policy until 1926 (or, legally, until 1931) and Canadian foreign policymakers displayed little enthusiasm for involvement in world affairs. The result was that although Canada was geographically part of the Western hemisphere, rarely did Canadians recognize this fact. In their 1938 overview of Canadian foreign policy, Robert MacKay and Benjamin Rogers admitted to geography's "major influence" on policymaking, and, in this regard, they pointed to Canada's position as "an Atlantic country, a North American country, and a Pacific country" (3–5). For these writers, who went on to stellar careers in the Canadian foreign service, labelling Canada as "North American" applied only to relations with the United States and did not mean that it was a country of the Americas, a region to which they devoted little attention, for Canadian interests there were "almost wholly commercial and financial" (MacKay and Rogers 1938, 145). As was so often the case, the United States obscured Canadians' view of the rest of the Americas, including Mexico, long excluded from Canadian definitions of North America.

Throughout the 1920s and 1930s, interest in Ottawa in greater involvement in the Western Hemisphere remained at a low ebb, though O.D. Skelton, the forward-looking Undersecretary of State for External Affairs (USSEA), did advocate for expanding Canada's diplomatic presence globally, including in Latin America. As with so many issues in Canadian foreign policy during this period, the isolationism of prime ministers R.B. Bennett and William Lyon Mackenzie King proved too strong (Hillmer 2015, 189, 260–1). Despite slowly growing Canadian diplomatic involvement in the interwar world, with missions established in Geneva, Washington, Paris, and Tokyo, thinking in Ottawa was dominated by Canada's growing ties to the United States and its vital links with Britain. Such links – economic, military, cultural, and constitutional – were strong enough, or were perceived in those terms, that successive US administrations, fearful that Canada was a stalking horse for British interests, consistently blocked Canadian membership in the PAU. On this point, there was agreement between Americans and at least some Canadians: Academic Frank Scott noted that disinterest in PAU membership in Canada stemmed from "the sense of being a

'British' country" (Scott 1938, 117; Anglin 1961, 1–20). Canadians had yet to view Canada in hemispheric terms.

During the Second World War, Ottawa began establishing diplomatic relations with states in the region: Brazil and Argentina (1940), Chile (1941), Mexico and Peru (1944), and Cuba (1945). This move was taken mainly because of the loss of trading partners in German-occupied Europe. The war, admitted diplomat Dana Wilgress, "appeared to offer an opportunity of extending our influence in the Western Hemisphere, south of the River Rio Grande." Travelling the region to drum up trade prospects, Wilgress recalled being warmly welcomed by locals who "interpreted our mission as a belated recognition of the fact that we lived in the same hemisphere" (Wilgress 1967, 116–18; Murray 1974, 153–72). For the few commentators on Canada's foreign policy, there was a sense that this geographic fact was slowly being recognized. "Officially," one writer noted in 1941, Canada "has scarcely acknowledged the existence of a world south of the Rio Grande ... a vast unknown," of little importance beyond matters of trade and investment (Corbett 1940–1, 781). The following year, Frank Scott observed that Canadians had "begun to think of ourselves as an American nation and to realize that we may have duties to fulfill in the Americas quite as pressing as any duties in Europe" (Scott 1942, 8–16). Yet, old habits died hard. Canada was North American, admitted journalist George Ferguson, "but its history, its traditions, its sentiments and prejudices and, above all, its important economic interests, make Canadians look east and west rather than south." As Ferguson concluded, for Canadians "there is no meaning in the term 'hemispheric relations'" (Ferguson 1943, 47–51). Exemplifying this sense of this term's limits, conservative politician (and future high commissioner in London) George Drew praised the concept of "Pan American cooperation," but added that in his view the way to its fulfilment was "an ever closer understanding between the English speaking people," that is, the bringing together of the United States and Britain via Canada (Drew 1941, 141–7). For Drew, and doubtless for many other Canadians, Pan-Americanism stopped with the United States. Furthermore, Drew's comments indicate the pull of the concept of the North Atlantic Triangle, which reached its apogee during the Second World War.

Expanding diplomatic ties in the 1940s and attendant increases in trade failed to lead to a greater Canadian political commitment to the region. Attention had focused on Canada's avoidance of the PAU and its successor body, the Organization of American States (OAS). In an era in which Canadian officials supported and joined a host of new multilateral bodies, studious aloofness from the hemisphere was telling, though, as noted, Washington had been wary of Canadian involvement. With the dawning Cold War and Canada's emergence from colony to independence, the US calculus changed. As Arthur Vandenberg, the influential Republican senator, declared in 1947, the time "may come soon when our Continental fellowship will be geographically and spiritually complete ... from the Arctic circle to Cape Horn" (Miller 1947–8, 24). However, officials in Ottawa were swayed little by such sentiments, for they feared being drawn into inter-American disputes (Ogelsby 1969, 571–89; McKenna 1995). Important, too, was that while Vandenberg was voicing his paeans to geographical and spiritual brotherhood, Vincent Massey had advanced a very different viewpoint. Massey's contention that Canada belonged to the Northern rather than the Western Hemisphere found wide support in post-war Ottawa, even as a sea change was taking place in Canada's foreign policy. When Mackenzie King left office in 1948, internationalist-minded officials – most notably Prime Minister Louis St. Laurent and Secretary of States for External Affairs (SSEA) Lester Pearson – came to power and pushed Canada out into the world.

Latin America in the "Golden Age" of Canadian Foreign Policy

Despite the post-war expansion of Canadian interest in multilateral fora and "enthusiastic participation in the details of international life," the Western hemisphere proved too far away (Holmes 1979, 4). Relations were established with Colombia, Venezuela, and Uruguay (1952), and the Dominican Republic and Haiti (1954), but otherwise Pearson, a man synonymous with Canadian internationalism, devoted little action to the hemisphere. Speaking on the theme of "Canada in the Americas," Pearson focused most of his attention on the United States, mentioning only briefly that Canadian officials were "looking south as well. In fact, we are looking in all directions" (Canada DEA 1947). Despite this speech's title, then, for Pearson the Americas were a small part of much larger, global picture, where what was coming to be called the West obscured Latin America. Commenting on the nascent North Atlantic Treaty Organization (NATO), Pearson specifically noted that in contrast with the PAU, the North Atlantic community "reflects political, economic and cultural interests which in the history of Canada have been of importance in the growth of its freedom and security" (Pearson 1948–9, 374).

In the early post-war years, official Canadian attention was transfixed on the North Atlantic, hardly a surprise given that Canada had helped to liberate Western Europe during the world war (the second in three decades) and that Ottawa was concerned by Soviet hostility. Even so, the reasons for Canadian interest in this region and not Latin America went beyond strategic calculations. A brief for the prime minister affirmed that Canada had a vital interest in defending Europe, from which "Canada has received nine-tenths of its people, both its languages, its religion, and most of its laws and political institutions." "The Canadian people," it continued, "unlike the peoples of the other nations of the Americas, have never ... turned their backs on Western Europe" (Western Union Survey 1948). "For us in North America," one influential diplomat stated, "the shrines of western Europe are no mere items of geography. In Britain, in France, in Italy are the vital wellsprings of our civilization" (Canada DEA 1951). Increasingly, too, as the Cold War shifted across the Pacific and as the post-war wave of decolonization unfolded, Asia came into the official Canadian picture. But not Latin America. From their analysis of Canada's place within the inter-American system and the lack of Canadian public knowledge about Latin America, in 1948 two commentators concluded: "The simple fact remains that most Canadians have, and will continue to have, for as long as one can foresee, more in common with countries of the British Commonwealth or with Scandinavians than with Hondurans or Paraguayans" (Soward and Macaulay 1948, 44).

It is telling of these sentiments and perceived Canadian interests that during his storied nine-year stint as SSEA (1948–57) Pearson avoided the region. For Canada, Latin America was important only in economic terms. In 1953, Trade and Commerce Minister C.D. Howe undertook a month-long goodwill visit to the region. "Like Canada," Howe remarked upon his return, "the countries of Latin America are countries of the future. We have come closer together over the years. May I express my sincere hope that we shall continue to work together in our programme for the expansion of our economies" (Canada DEA 1953). This was a fine sentiment, and Howe's department acted upon it. Still, the political limitations on Canadian involvement were clear. In the wake of Howe's visit, diplomats explored whether his trip had created any groundswell of support for diplomatic initiatives in the region, including OAS membership. USSEA Wilgress judged that despite Howe's mission, there remained a dearth of public support for such moves, a thankful development, he added, because there was a section of the public who would react negatively to any initiatives in Latin America out of a sense that they were "another step in the direction of weakening our ties with the Commonwealth" (Wilgress 1953). The following year, during a Cabinet discussion over how Canada should respond to a US query about whether Ottawa wished to attend an OAS economic conference, one minister observed that "there were feelings in some quarters that the government was not as concerned with this organization as it was with others composed in large measure of English speaking, Protestant nations" (Cabinet Conclusions Memo 1954). In short, during the St. Laurent-Pearson years, continued British imperial feeling and an Atlanticist outlook were roadblocks to engagement with Latin America.

British sentiment was a mainstay of the world view of Progressive Conservative John Diefenbaker, who became prime minister in 1957. In addition to rekindling Anglo-Canadian ties, he was keen to expand Canada's global presence, moves aimed at diversifying Canada's economy away from increasing reliance on the United States. Furthermore, his time in office corresponded with mounting decolonization, which underscored the need to branch outwards. In terms of Latin America, Diefenbaker took power just as the region saw the collapse of a raft of authoritarian governments, notably in Argentina, Brazil, Colombia, and Venezuela. The Tories reached out to these fragile democracies: Canada further expanded its presence, establishing diplomatic relations with Ecuador, Bolivia, Paraguay, and the Central American states; within the Department of External Affairs (DEA), a specific Latin America Division was formed; Ottawa hosted a string of political leaders from the region, and in 1960 Diefenbaker visited Mexico, becoming the first Canadian prime minister to officially travel to Latin America. There were visits, too, by his two foreign ministers, Sidney Smith in 1958 and Howard Green in 1960. Moreover, Canadian officials began to attend OAS meetings as unofficial observers, and OAS membership was seriously contemplated (McKercher 2012, 57–80).

These developments helped increase Canadian awareness of the hemisphere. For Smith, his trip to Mexico and Brazil expanded his own mental map, for, as he told fellow parliamentarians, "I learned, not only some geography but also some psychology … It just seems to be inherent that they [Latin Americans] do not think of the American hemisphere as being divided into two continents" (Canada House of Commons 1959, 8–9). Howard Green, fresh from visiting Argentina, Chile, Peru, and Mexico, explained that the Latin Americans he had encountered "are deeply puzzled that Canadians do not seem to realize that Canada is a very important member of the Western Hemisphere family." In announcing the reorganization of the DEA to take greater account of Latin America, he affirmed that the move reflected a realization that "many western hemisphere problems involve us … and we cannot get away from the effect of many of the events which take place on this side of the Atlantic" (Canada House of Commons 1960a; 1960b).

For some commentators, these moves by the Diefenbaker government were a welcome sign that it recognized Canada's geographic position. "Canada finally joins the Western Hemisphere," one columnist wrote approvingly. Charles Lynch, another journalist, disagreed, noting in his columns as well as in a public address, that it was OAS membership that "would make us officially part of the Western Hemisphere" (Sclanders 1960; *Montreal Star* 1961).

Viewed as a barometer of Canadian engagement with the Western hemisphere, the OAS membership issue emerged from Diefenbaker's increasing attention to Latin America. Ultimately, he resolved to avoid the organization, in part because visiting Ottawa in 1961, John F. Kennedy urged joining up and the Canadian prime minister did not wish to appear cowed by the US president. Furthermore, Canada's government was mindful of the spread of Cold War politics to Latin America: Cuba's clash with the United States and its increasing isolation from the rest of the hemisphere – the OAS expelled Cuba in 1962 and regional governments broke diplomatic and trade relations – raised regional tensions. Ottawa's decision to maintain economic and diplomatic ties with Havana made Canada a hemispheric outlier (Kirk and McKenna 1997). Finally, the Diefenbaker years saw the start of the cycle of revolutionary and counter-revolutionary violence that gripped Latin America and that lasted until the 1990s. Together, these were reasons enough to steer clear of the region, one where Diefenbaker's prime interest was in expanding Canadian economic prospects, a reminder that, when it comes to Latin America, Canadian governments of all stripes have shared this singular goal (McKercher 2018).

Beyond the OAS issue, which was rarely a front-page issue in Canada, it was the Cuban upheaval that drove increasing Canadian attention to Latin America. "I believe there is a much greater awareness ... now," Green told a House of Commons committee in 1961, "and that has been increased by the troubles in Cuba" (Canada House of Commons 1961, 67–8). "With Castro's taunting of the United States, Canada is now well aware of Latin America," wrote one journalist in 1960, as the US trade embargo went into effect. "But it is only very recently that we have shown such interest" (Knowles 1960). And contending that Canada should join the OAS – a rare stance among Canada's newspapers – the *Toronto Star* editorialized in 1961 that Canadians could no longer "pretend that Latin America can safely be excepted from the world-wide concerns and responsibilities that this country has assumed. The world has shrunk to the dimensions of a neighborhood, and Latin America's is the next backyard but one from our own" (*Toronto Star* Editorial 1961). For some Canadians at least, the geopolitical upheavals of the 1960s placed Latin America increasingly into focus, though involvement in these troubles was another issue entirely.

Despite growing awareness of Latin America, the limiting factors of geography and identity remained. Fresh off his trip to Mexico, Diefenbaker complained that "Canadians seldom look beyond the United States," and while the prime minister likely did not need the reminder, USSEA Norman Robertson pointed out to him that "Canadians have traditionally thought in terms of the North Atlantic Community and Commonwealth associations, and are not on the whole attracted to the concept of hemispheric solidarity" (Diefenbaker to Green 1960; Robertson to Diefenbaker 1961). This theme was taken up by John Holmes, a former golden-age diplomat who did much, both within government and outside of it, to promote the notion of Canada as an engaged middle power. Holmes was no fan of concepts like hemispheric solidarity. He was keen to stress, as he did to an international conference on Latin American tensions in 1962, that, "Hemispheres are, after all, figments of the geographers' imagination" (Holmes 1962, 415). Dismissive of what he saw as emotional appeals to Canadian political involvement in Latin America, in 1967 he admitted to geographic realities, writing that "Canada is a part of the Northern as well as the Western Hemisphere," but added that he saw little reason to assume that geography gave Canadians a "special affinity" with either Mongolia or Paraguay. Thinking in hemispheric terms, for him, was "distorted geography," and while there might be reasons of national interest for "why Canadians should help Bolivians as well as Nigerians ... it is hard to understand why there should be a priority for Bolivia on geographical grounds" (Holmes 1968, 173–6, 183). Both Robertson and Holmes were of the same generation of Canadian diplomats as Pearson, and together they shared an outlook – hardly an outlier – that limited Canada's place in the Western hemisphere.

Yet when Pearson succeeded Diefenbaker as prime minister in 1963, there was some expectation that the new Liberal government would bring Canada into the OAS. Not only had Pearson mused during the election campaign about taking this step, but Paul Martin Sr., his foreign minister, was a longtime proponent of outreach to the Western hemisphere, declaring in 1936: "our region is America" (Martin 1936, 119). Three decades later, Martin remained convinced that joining the OAS was the "ultimate destiny of Canada as a country of the Western Hemisphere" (McKercher 2014, 480; Dirks 2017). This same view was publicly advanced in 1965 by Arthur Irwin, who had recently retired as the Canadian ambassador to Mexico. Irwin's essay advocating OAS membership was based on a private report that he had authored for

Martin's DEA. As he explained, Canada was an "American nation," sharing with Latin American countries not only common cultural and spiritual influences found in a shared European heritage, but also "the travail of pioneering a new world, of shaping political, economic and social institutions to a new world environment, and of wresting by one means or another political independence from colonial status." Beyond these cultural issues, membership, he concluded, would confirm the "geographical logic" of the Americas. In his private report to Martin, Irwin added that the need to branch southwards was reinforced by the fact that "the Atlantic-European community to which we belong had lost dominance as a prime mover of world history," and so Canada had best expand its horizons (Irwin 1965, 294–5; Irwin Memo 1964). Here was an effort to convey common geographic and cultural links between Canada and Latin America. Irwin's public conclusions generated push-back. Detractors pointed to the increasing Cold War atmosphere in inter-American relations – the United States had invaded the Dominican Republic in 1965 and a string of right-wing coups had begun toppling the region's democratic governments – as a cause for caution. Furthermore, there were objections to Irwin's mental cartography. One political scientist contended that "while the logic of map makers might be served [by OAS membership], it is hard to see who else would benefit." "Canada does not belong within the OAS," added the *Globe and Mail*, for it was vital to respect "the traditional bonds that still orient the Canadian outlook toward Northern Europe" (Smith 1966; *Globe and Mail* Editorial 1965).

In the end, the Pearson government, like its predecessor, spurned OAS membership, an issue that had lurked prominently in discussions about Canadian foreign policy towards Latin America for almost a decade. Reflecting in 1969 on the issue, Jack Ogelsby, Canada's leading commentator on inter-American affairs at the time, noted that, if nothing else, the question had "focussed Canadian attention on Latin America. That is, perhaps, its greatest contribution" (Ogelsby 1969, 589). Indeed, Pearson himself admitted to a change in his own mental map. Commenting on the OAS issue in a press conference in 1965, he explained that Canada had been slow to recognize "that we are a Western Hemisphere country," with Canadians' focus "directed across the Atlantic rather than down into South America," adding that such a view was becoming outmoded (Pearson 1965). Several months later, he told the Argentine ambassador that "hitherto Canada's background had led it to look primarily to Europe but we had been realizing more and more the importance of our relations with Latin America and with the countries across the Pacific" (Memo to Cadieux 1966). In these comments, we see Pearson's sense of Canadians' shifting mental maps. Certainly, the growing importance of events in the Global South throughout the 1960s focused Canadian attention on the world beyond the North Atlantic. It was this vast area of the globe that became a major focus for Pearson's successor, Pierre Trudeau. An intellectual who prided himself on breaking taboos, Trudeau challenged – at least initially – much of the status quo in Canada's foreign policy.

The Trudeau Pivot to Latin America

As part of Trudeau's effort to shift the focus of Canadian foreign relations, Latin America assumed greater importance in Ottawa, where thinking in hemispheric terms became more frequent. In May 1968, in his first speech on foreign policy as prime minister, Trudeau spoke of the need "to take greater account of the ties which bind us to other nations in this hemisphere" (Canada DEA 1968). Later that year, five of his ministers, the heads of ten government agencies, and a smattering of academics and journalists spent a month travelling to Argentina, Brazil, Chile, Colombia, Costa Rica, Guatemala, Mexico, Peru, and Venezuela. Unprecedented on Canada's part in terms of outreach to Latin America, this ministerial visit emphasized the placement of this region within Canadian mental geography. This "'voyage of exploration,'" a briefing note explained, aimed to give the new government "a clearer conception of the place which Latin America as a whole – and the countries of Latin America considered separately – should occupy within the framework of our foreign relations" (Ministerial Report 1968). In the resulting report on their mission, the ministers outlined a policy of engagement with the Western hemisphere and blamed past tepidness on "distance ... the fact that Canada is geographically separated from Latin America by the United States and ... historical pre-occupations" with Europe. In terms of the latter factor, the report praised the fact that "like the people of Latin America the great majority of Canadians are the heirs of the Christian traditions and of the Greco-Latin civilization," thereby underlining the importance of cultural links that had been largely denied or obscured in previous analyses (Canada DEA 1969a). Together, these early developments indicate the new government's effort to take account of Canada's place within the Western hemisphere, even if it signalled its limited understanding of the Indigenous or Afro-descendent peoples in that region.

In the wake of the ministerial visit, an inter-departmental task force chaired by the DEA considered future relations with Latin America, part of the Trudeau government's commitment to produce a new foreign policy for Canadians. Their resulting study highlighted the negative factors of Canadians' "evanescent" awareness of the region thanks to upheavals of the 1960s and the continual problem posed by the United States, which stood as a barrier between Canada and Latin America. However, the study went on to predict that throughout the coming decade, expanding trade would boost familiarity with the region. Furthermore, globalization – what the study referred to as the "revolution in transport and communications" – was diminishing "Canadians' sense of isolation" from the hemisphere. The task force's report was careful to state that increasing trade, tourism, and media attention would not necessarily mean that "Canadians will feel psychologically as close to Latin America as to Europe or the United States; Tierra del Fuego, for a Canadian who has heard of it, will still seem a lot farther away than the coast of Brittany or the Grand Canyon." Nonetheless, it concluded, "there are now hundreds of thousands of Canadians for whom Acapulco is more real than Edinburgh or San Francisco" (Canada DEA 1969b). Such observations and predictions are revealing of how the process of globalization, which accelerated throughout the 1970s, shrunk space and time and began to alter perceptions of the world. The expansion of Canadian tourism to Latin America (or at least the circum-Caribbean) certainly increased awareness of the region, even if in a superficial manner.

In terms of Canadian foreign policy, the expanding mental map that situated Canada within the Western hemisphere increased attention towards this region. In *Foreign Policy for Canadians* (1970), the Trudeau government's five-part outline of a new approach to the world, Latin America received its own volume, which pointed to changing mental maps in government circles. "Canadians, as they look abroad," it stated, "are increasingly conscious that Canada is a distinctive North American country firmly rooted in the western hemisphere." Yet however much Canadians might be beginning to see themselves in hemispheric terms, the document highlighted the presence of the United States, which "screens Canada from Latin America," and would continue to "condition Canada's relations with the area south of the Rio Grande" (Information Canada 1970, 5–6).

In practical terms, the increased attention to Latin American translated into the extension, in 1970, of official development assistance (ODA) to the region via bilateral development programmes. Up to that point, the Canadian International Development Agency (CIDA) and its proto-predecessor had been focused on the British Commonwealth and Francophone Africa, though in 1964, Canada had begun contributing a small pool of funds to the Inter-American Development Bank

(IDB). Canada officially joined the IDB in 1972 and boosted its contributions. Throughout the 1970s, and under the leadership of CIDA President Paul Gérin-Lajoie, spending on development aid in Latin America
increased exponentially, and even came to encompass Cuba as part of
Trudeau's efforts to bridge the region's Cold War divide (Gérin-Lajoie
1976). Also, the period 1970–6 saw four ministerial trade missions as
well as a tour that brought the prime minister to Mexico, Cuba, and
Venezuela. Furthermore, Canada joined both the Pan American Health
Organization (1971) and the Inter-American Institute of Agricultural
Sciences (1972), and in 1972 Canada became a permanent observer to
the OAS, although it refrained from formal membership. These moves
were in line with the Trudeau government's general expansion of relations with the Global South. Even so, there were limits to new geospatial thinking in that Latin America hardly took pride of place in what
was a global foreign policy on Trudeau's part.

Despite these Trudeau-era advances, observers of Canada–Latin
America relations saw a disconnect between official thinking and public opinion. As late as 1976, Jack Ogelsby observed that "Canadians,
an American people, have not, as a whole, been very interested in the
America that stretches from Cape Horn to the Arctic Islands." In an
observation that continued resonating over the next four decades, he
added that Canadians "have even appeared to deny Mexico its place in
North America," preferring, instead, to limit the continent to Canada
and the United States (Ogelsby 1976, 1). That same year, John Harbron,
another longtime observer of and advocate for links with Latin America, similarly lamented that most Canadians "preferred to remain more
European than American" and to emphasize ties to the British Commonwealth and la Francophonie, "two global commonwealths, neither
indigenous to this hemisphere" (Harbron 1976, 109). The change to
mental maps was a slow process, what Jean-Pierre Goyer, the parliamentary secretary for external affairs, likened in 1970 to a process "of
mutual discovery" on the part of both Canadians and Latin Americans.
What Goyer himself had discovered was that "in many ways Canada's
heritage is similar to that of Latin America. Each is a vast territory, rich
in natural resources, much of it undeveloped or even unexplored. Our
populations are mainly European in origin, with varying admixtures of
indigenous peoples and of immigrants from other parts of the globe"
(Canada DEA 1970). Echoing other advocates of closer Canada–Latin
America ties, such as Arthur Irwin, Goyer was keen to point out similarities among Canadians and Latin Americans, all in an effort to make
the hemisphere less foreign.

What Goyer, Gérin-Lajoie, and Trudeau all shared was their French-
Canadian heritage, and as Ogelsby observed, the Trudeau government's

attention to Latin America sprang not only from interest in the Global South but from francophones' "sympathy with the concept of *latinité*" (Ogelsby 1979, 187). If the Anglocentrism of Massey, Diefenbaker, and Pearson limited their interest in the Western Hemisphere, then common links between French Canadians and Latin Americans generated at least some sense of sympathy, interest, and common understanding. Certainly, that is how the situation appeared to various observers. Writing in 1948, two commentators noted: "Sentimental arguments about the special affinity of Canada with Latin countries, which has some appeal in Quebec, need not be taken too seriously" (Soward and Macaulay 1948, 43). That same year, in assessing Pan-American sentiment among francophones, Iris Podea wrote that since Latin American culture continued to thrive despite US power, reaching out to the region could provide francophones with a reinforcement of their "way of life in the face of an overwhelming Anglo-Saxon environment." For Podea, ties between French Canada and Latin America were important if "largely psychological" (Podea 1948, 335–48).

To flash forward a decade, when Fidel Castro visited Montreal in 1959, one journalist described it as: "a Latin visiting with other Latins," and two years later a newspaper writer referred to Latin America as this "other Latin community, which we must not exaggerate of course, but no longer ignore." For *Le Devoir* editor Claude Ryan, French Canadians had a "special cultural affinity" for Latin America (Ayotte 1959; Léger 1961; Ryan 1964). Given these affinities, perhaps, there was less reticence on the part of French Canadians in advocating for stronger links with Latin America. Professor Marcel Roussin felt that Canada, "as an American nation, has the sacred and undeniable duty to participate to the full extent of its forces in the defense and development of the continent where Providence has placed it," and Pierre Trudeau's own *Cité Libre* advocated for OAS membership "because Canada was an American nation" (Roussin 1959, 253; 1964, 2–3). Importantly, many of the Canadian diplomats sent to Latin America during the 1950s and 1960s were francophone, as was Paul Martin. However, in an analysis of Canadian public interest in Latin America from 1957 to 1967, commissioned by Martin's DEA, Jack Ogelsby found little difference between English and French-Canadian opinion on hemispheric issues. "But the point," he wrote, "has been made that almost everyone, English or French, *thinks* there is this affinity, and it is ever regarded thus" (Ogelsby 1976, 328; see also Gay 1983).

With his focus largely on press opinion, Ogelsby overlooked missionary work, much of it conducted by French-Canadian Catholics, a major element of Canadian involvement in Latin America, one that long preceded interest by Canada's government. Religious and cultural diplomacy formed a key link between Quebec and Latin America, generating sympathy and understanding. For instance, the leader of the Montreal-based Union des Latins d'Amérique, a cultural association, declared in August 1944 that, in Latin America, "French Canadians have found spiritual brothers, people influenced by the same Catholic and Latin civilisations as ours" (Demers 2016, 197). Bolstered by such conceptions, many missionaries worked within a transnational solidarity network linking Latin America and Quebec in a common nationalist struggle (LeGrand 2009, 43–66; Gauthier and Lord 2013; Demers 2014).

This sense of a shared struggle was particularly strong from the 1960s to the 1980s, a period of intensifying Québécois nationalism alongside Latin America's revolutionary upheavals. For instance, Pierre Vallières, a radical Quebec nationalist, wrote that "The Québécois are also Latin Americans and, like the peoples of South America, are part of the 'Third World.'" Going beyond emphasizing cultural affinities, Vallières had situated Quebec within the Global South, a neat trick of geospatial thinking that underscored the salience of Québécois nationalists' struggle "against imperialism, for national liberation" (Vallières 1967; LeGrand 2013, 93–116). Viewing Quebec as a Third World country placed the province within a revolutionary mental map. Along similar lines, Québécois nationalists viewed the 1973 Chilean coup as an example of the ways in which Anglo-Saxon capitalists stomped upon Latin workers, whether in Chile or in Quebec. Commenting on the overthrow of Salvador Allende's government and anti-union activities by mining firms in northern Quebec, one activist reflected: "Chileans and Québécois simultaneously felt the blows to a force that wanted only to take into account the needs of the people rather than the voracious interests of multinational corporations" (Mongeau 1974). Given Québécois' sympathy for the short-lived Allende government, Quebec became a major recipient of the more than 7,000 Chilean refugees who fled to Canada following the coup. These Chileans, and thousands more from other Southern Cone countries, were allowed into Canada only after the Trudeau government reluctantly succumbed to considerable public pressure by activist and solidarity groups (del Pozo 2009; Peddie 2014).

Making a Claim of Belonging

As John Foster's chapter in this collection makes clear, the 1960s and 1970s witnessed the growth of Latin American solidarity movements throughout Canada. First, the Cuban revolution and then this period's political turbulence brought the hemisphere into increasing focus, particularly for leftists and human rights activists interested in advancing social justice, protecting refugees, and punishing the region's odious military regimes (Sheinin 2009; Wright 2009). As in Quebec, many of these early solidarity activists drew comparisons between their own struggles – against capitalism and US imperialism – and conflicts in Latin America (Lumsden 1968). One activist took aim at multinational corporations – some of them Canadian – and criticized "The violence of the owners of the smelters at Sudbury, of Brazilian Light or the Bank of Nova Scotia, engaged in exploiting Latin America or the Canadian hinterland" (Resnick 1970, 96). Similarly, the Toronto-based Latin American Working Group noted that "Canadians are exploited and share in exploitation," for while Canadian companies were complicit in Chile, "Canada itself is in many respects confronted with the same problem as Chile, seeking ways to develop an independent society in the face of external corporate domination" (LAWG 1973; see also Swift 1977; Nef 1978). These solidarity networks were just one sign of growing transnational links that placed the Western Hemisphere more firmly onto the mental maps of at least some Canadians. As John Harbron observed in 1974, while Canada had long "denied its existence as a nation of the western hemisphere," incidents like the Chilean coup "have ended the tendency to isolation. For better or for worse, Canada will not again deny its geographic place in its own hemisphere" (Harbron 1974). Solidarity networks strengthened in the 1980s following the intensification of violence in Central America. Working both in Canada and in Nicaragua, Guatemala, and El Salvador, Canadian activists rallied to protect Central Americans, to challenge US policy and the violence of reactionary military and paramilitary forces, and to urge Canada's government to stand up for human rights and take in refugees (McFarlane 1989; North 1990; Anderson 2003). Whether by finding common ground on social justice issues or working to help victims of violence, the growth of transnational activism in Canada in the 1970s and 1980s showcased one way in which mental maps were expanding.

The Trudeau years saw the formation of other transnational ties between Canada and Latin America. One element that distinguished Trudeau-era foreign policy from previous governments – setting a pattern for the future – was the extent to which officials partnered or consulted with non-governmental organizations (NGOs) (Stevenson 2000). In this respect, the Canadian Association for Latin America (CALA) was a prominent, business-friendly group formed in 1969 to promote corporate-led development. Receiving CIDA funding and support, CALA's director stated that his organization's primary goal was to break down Canadian ignorance about Latin America and combat "basic misconceptions about Latin American politics and temperament"

(Lubbock 1976, 19–23). Doing so would be no easy task, for as a CALA-commissioned study pointed out, the United States blocked Canadians' view of the hemisphere and, worse, there "does not seem to be a geographical rationale justifying a major thrust towards Latin America. Canada's Atlantic, Pacific, and Francophile interests are much deeper" (Bradford and Pestieau 1971, 73). Clearly, combating ignorance of the region was important, as was emphasizing Latin America's consequence to Canada irrespective of geography. Happily, then, 1969 also saw the formation of the Canadian Association of Latin American Studies (CALAS), a scholarly association later to become the Canadian Association of Latin American and Caribbean Studies (CALACS). Its founding signalled the growing number of Canadian scholars interested in Latin America. The 1964 *International Studies in Canadian Universities* report had noted the paucity of attention given to Latin America in Canada's universities. Yet in the inaugural issue of CALAS's journal *North-South*, scholar Walter Soderlund pointed to the "phenomenal increase in the number of Canadian universities offering courses in the Latin American area" (Soderlund 1976, 2). Overall, then, the Trudeau years saw increasing Canadian attention to and awareness of Latin America, whether in terms of civil society groups, activist networks, or government officials.

Perhaps given the uptick in hemispheric thinking, it is no surprise that Brian Mulroney's government, which took power in 1984, had no compunctions about situating Canada in hemispheric terms. "We are a nation of the Americas," it noted in its 1985 foreign policy white paper, "with an interest and an investment in the hemisphere's future" (Canada Supply and Services 1985, 1). Speaking to the Inter-American Press Association in Vancouver the following year, Mulroney praised a shared "vital and vested interest in the advancement of hemispheric relations." "Wherever we come from in the Americas," he added, "however we came to the Americas, we have elements of a common heritage of geography, history and culture" (Canada DEA 1986). In October 1989, Mulroney visited the meeting of the OAS in San José, Costa Rica, which marked his first visit to Latin America. Here, he announced that Canada would enter the organization, a sign, for one observer, that Canada was "becoming a country of the Americas in fact rather than in rhetoric" (Dosman 1989, 1). Privately, the prime minister reflected that pressing issues such as narcotics control and the environment were influenced by events "in our own hemisphere" and it was time to "play a constructive role in our own backyard" (Mulroney 2007, 692). Long in the making, OAS membership was a concrete development, but still there was a sense that it was out ahead of public attitudes. In remarks delivered during the ceremony in which he signed on to the

OAS Charter, SSEA Joe Clark reflected on the "psychological distance between Canada and the southern part of this hemisphere. We intend to change that state of mind, to make Canada 'exist' more significantly for Latin America and to make Latin America 'exist' more substantially for Canadians" (Canada DEA 1989). Nor was Clark alone in sensing the limits of Canadian mental maps. Edgar Dosman, founding director of FOCAL, the Canadian Foundation for the Americas, a think tank created in 1990, reflected on Mulroney-era progress, counselling that there remained "a certain fragility" to Canada's dealings with Latin America, "as if the historic lack of a long-term political commitment still impedes an irreversible reorientation of Canada's relations with the Americas" (Dosman 1992, 529).

Entering the OAS in 1990 was a milestone, not a stopping point. Once a member of the organization, Canada took part in efforts promoting democracy and post-conflict stability in the region. Several of these efforts had begun under UN auspices – participation, for instance, in the UN Observer Mission to Verify the Electoral Process in Nicaragua – but involvement only accelerated after 1990. Furthermore, there was a flurry of trade promotion, a reminder that while the Mulroney government negotiated the 1988 Canada–US Free Trade Agreement, its economic horizons extended beyond its massive southern neighbour. Indeed, Mulroney agreed to roll the bilateral trade agreement with the United States into a trilateral deal with Mexico. NAFTA marked a recognition – one that waxed and waned – that Mexico was a part of North America rather than apart from it. Thus, although Jean Chrétien had campaigned against NAFTA, once in power in 1993 he backtracked. In trade terms, the Chrétien government looked beyond North America towards the Americas writ large: Ottawa backed the ultimately unsuccessful Free Trade Agreement of the Americas process. Whether as a reflection of his French-Canadian heritage or a sign of his past as a Trudeau cabinet minister or recognition of Latin America's changing place in elite mental maps – or a combination of the three – Chrétien himself embraced a hemispheric identity. "Here, in Santiago," he told fellow delegates to the Second Summit of the Americas in 1998, "it is clear that we are becoming something more than amigos. We are *una gran familia*." Using the same phrase while addressing Mexico's Senate the following year, he added: "from Baffin Island to Tierra del Fuego, our hemisphere is becoming not just a group of nations connected by an accident of geography ... But by an active identity" (Chrétien 1998; Chrétien Excerpts 1999, 13). In his memoir of his time as Chrétien's foreign minister (2002–4), Bill Graham recounted that he had wanted to prioritize Latin America. "We still tended to see ourselves as a North

Atlantic nation," he reflected, "midway between Europe and the United States. We were beginning to see ourselves as a Pacific nation as well. Now we needed to integrate Canada into the Americas" via trade links and cultural exchanges (Graham 2016, 238). Mental maps were changing, though towards the end of Chrétien's premiership, Canadian focus shifted towards Afghanistan.

Conclusions: Interpreting the Present

What, then, of the more recent past? In 2007, a FOCAL-commissioned public opinion survey found that "Canadians do not feel that Canada is similar to many Latin American countries" (FOCAL 2007, 7). Five years later, one foreign policy analyst wrote that the Americas were "fully embedded in Canadian diplomatic thinking and practice" and had "moved into the geographical and functional mainstream in terms of both identity and interests" (Cooper 2012, 687). Apparent in these observations is an important divide between elite and public opinion, that is between where Canadian government officials situate Canada and where Canadians writ large see the country. In 2009, Stephen Harper's government launched its "Americas Strategy." Continuing a trend that extended back to the Trudeau years, the strategy document recognized that "Canada is a country of the Americas" (Canada DFAIT 2009, 4). Yet many Canadians seem to have continued to ignore this fact even as the forces that drove much Canadian attention to Latin America beginning in the late 1960s – tourism, trade, media attention – have only accelerated in importance and as the flow of immigrants and international students from the region to Canada has increased considerably. Canada has become a country that accepts immigrants from all over the world, and so in terms of Canadians' globalized identity, Latin America is part of a large picture that perhaps obscures the region. A related issue here is Canada's own place in Latin American mental maps. "The country needs to become known in the hemisphere," noted Carlo Dade, for "while the Americas are becoming increasingly important for Canada, the same can't be said the other way around" (Dade 2011).

As for the Americas Strategy, rhetoric aside, the Harper government focused on trade and investment, indicating that, policy-wise, little has changed over the past century. As one analyst of Harper's approach put it: "There's still Much to Discover" (Rochlin 2012; Macdonald 2016, 1–17). The same could be said of Justin Trudeau's Liberal government, which has done little to put Canada on the map as far as breaking away from his predecessor's Latin American policies. The one exception is the intensification of democracy-promotion efforts via Ottawa's instrumental role in the Lima Group's confrontation with the Nicolás Maduro regime. The Trudeau government's intense interest in Venezuela is an extension of the expansion of Canadian mental maps that took place from the 1960s onwards. As then foreign minister Chrystia Freeland noted in welcoming Lima Group members to Toronto in 2017, the confrontation with Maduro was important for "our region" (Canada 2017).

What the above review of Canada–Latin America relations has made clear, is that Canadians have been in a constant state of discovering the Americas and of realizing and recognizing Canada's place in the hemisphere. For much of Canada's history, cultural and geographic thinking led Canadian foreign policymakers to ignore Latin America. Slowly, through growing contact and changing cultural patterns, this thinking changed. Moreover, for a growing number of activists, Latin America began to figure prominently, leading to the growth of transnational ties. Indeed, this tendency was present, too, among Indigenous actors, who began to view the struggle for Indigenous rights in hemispheric terms. From George Manuel's 1970s conceptualization of a *Fourth World* occupied by Indigenous peoples throughout the globe, to Assembly of First Nations National Chief Ovide Mercredi's 1994 visit to Chiapas, Mexico, to efforts to secure passage of the 2007 UN Declarations on the Rights of Indigenous Peoples, Indigenous activists in Canada have found common cause with Indigenous people throughout the hemisphere. Still, however much Canada's place within the Western Hemisphere has come to be an accepted fact both in official Ottawa and in certain civil society circles, this region continues to occupy a limited space in Canadian mental maps. As the other contributors to this volume consider Canada's future place in the Americas, it is worth recalling the past, if only to be aware of the limitations imposed by geography and culture.

NOTE

1 Although referring to the Western hemisphere, I purposely ignore the British West Indies, which have their own supposed "special relationship" with Canada; see Brian Douglas Tennyson, ed. 1988. *Canada and the Commonwealth Caribbean* (London: University Press of America).

REFERENCES

Anderson, Benedict. 1991. *Imagined Communities: Reflections on the Origin and Spread of Nationalism*. London: Verso.

Anderson, Kathryn. 2003. *Weaving Relationships: Canada-Guatemala Solidarity.* Waterloo: Wilfrid Laurier University Press.

Anglin, Douglas. 1961. "United States Opposition to Canadian Membership in the Pan American Union: A Canadian View." *International Organization* 15.

Armstrong, Christopher, and H.V. Nelles. 1988. *Southern Exposure: Canadian Promoters in Latin America and the Caribbean, 1896–1930.* Toronto: University of Toronto Press.

Ayotte, Alfred. "Un Latin en visite chez d'autres Latins." *La Presse*, 25 April 1959.

Barney, Timothy. 2015. *Mapping the Cold War: Cartography and the Framing of America's International Power.* Chapel Hill: University of North Carolina Press.

Berger, Carl. 1966. "The True North Strong and Free." In *Nationalism in Canada*, edited by Peter Russell. Toronto: McGraw-Hill.

Bothwell, Robert. 2007. *Alliance and Illusion: Canada and the World, 1945–1984.* Vancouver: University of British Columbia Press.

Bradford, Colin, and Caroline Pestieau. 1971. *Canada and Latin America: The Potential for Partnership.* Toronto: CALA.

Braz, Albert. 2016. "Canada's Hemisphere: Canadian Culture and the Question of Continental Identity." *American Review of Canadian Studies* 46.

Brebner, J.B. 1945. *North Atlantic Triangle: The Interplay of Canada, the United States and Great Britain.* New Haven, CT: Yale University Press.

Broadhead, Lee-Anne. 2010. "Canadian Sovereignty versus Northern Security: The Case for Updating our Mental Map of the Arctic." *International Journal* 65.

Buckner, Phillip, and R. Douglas Francis, eds. 2006. *Canada and the British World: Culture, Migration and Identity.* Vancouver: University of British Columbia Press.

Canada. 1970. *Latin America: Foreign Policy for Canadians.* Ottawa: Information Canada.

– 1985. *Competitiveness and Security: Directions for Canada's International Relations.* Ottawa: Minister of Supply and Services.

Canada, Cabinet Conclusions. 17 September 1954, Library and Archives Canada [LAC], RG 2, vol. 2656.

Canada, Department of External Affairs (DEA). *Statements & Speeches (S&S)* 47/7, 8 March 1947.

– *S&S* 51/11, 19 March 1951.

– *S&S* 53/9, 26 February 1953.

– *S&S* 68/17, 29 May 1968.

– *S&S* 70/10, 30 June 1970.

– 1969a. *Preliminary Report of the Ministerial Mission to Latin America, October 27–November 27 1968.* Ottawa: Department of External Affairs.

- 1969b. *Report of the Latin American Task Force.* LAC, Mitchell Sharp Papers, vol. 90, file Foreign Policy in the Seventies.
- *Statement*, 15 September 1986.
- *S&S*, 89/62, 13 November 1989.
Canada, DFAIT. 2009. *Canada and the Americas: Priorities and Progress.* Ottawa: DFAIT.
Canada, GAC. Address by the Honourable Chrystia Freeland, for the third ministerial meeting of the Lima Group, 26 October 2017. Accessed 19 October 2021. https://www.canada.ca/en/global-affairs/news/2017/10 /address_by_the_honourablechrystiafreelandministerofforeign affair.html.
Canada, House of Commons. *Standing Committee on External Affairs* (5 March 1959).
- *Debates*, 30 May (1960a), 4335–8 and 14 July (1960b), 6298.
- *Standing Committee on External Affairs*, 2 May 1961.
Casey, Steven, and Jonathan Wright. 2008. *Mental Maps in the Era of Two World Wars.* Basingstoke: Palgrave.
Cavell, Janice. 2002. "The Second Frontier: The North in English-Canadian Historical Writing." *Canadian Historical Review* 83.
Chrétien, Jean. 1998. "Speech at Closing Ceremony of the Second Summit of the Americas." 19 April.
- Excerpts from Prime Minister Jean Chrétien's Speech to the Senate of the United Mexican States. 1999. *Canada World View* 4, 9 April.
Cooper, Andrew F. 2012. "Canada's Engagement with the Americas in Comparative Perspective." *International Journal* 67.
Corbett, P.E. 1940–1. "Canada in the Western Hemisphere." *Foreign Affairs* 19.
Dade, Carlo. 2011. "Latin America: The Glass Is Half …," *Globe and Mail*, 17 August.
Demers, Maurice. 2016. "Promoting a Different Type of North-South Interactions: Québécois Cultural and Religious Paradiplomacy with Latin America." *American Review of Canadian Studies* 46.
- 2014. *Connected Struggles: Catholics, Nationalists, and Transnational Relations between Mexico and Quebec, 1917–1925.* Montréal: McGill-Queen's University Press.
Diefenbaker to Green, 24 April 1960, LAC, RG 25, file 12426–40.
Dirks, John. 2017. "Friendly Noises but Distant Neighbours: Pearson, Latin America, and the Caribbean." In *Mike's World: Lester B. Pearson and Canadian External Affairs*, edited by Asa McKercher and Galen Roger Perras. Vancouver: University of British Columbia Press
Dosman, Edgar. 1989. "Mulroney's Discovery of the Americas." *International Perspectives* 18.
- 1992. "Canada and Latin America: The New Look." *International Journal* 47.

Drew, George. 1941. "A Canadian's View of Pan-Americanism." *American Association of School Administration Official Report 1941.*

Editorial. 1961. "Let Us Accept JFK Challenge." *Toronto Star,* 18 May.

Editorial. 1965. "Canada Does Not Belong within the OAS." *Globe and Mail,* 6 May.

Ferguson, G.V. 1943. "A Canadian's View of Hemispheric Relations." *Hemispheric Policy.* Washington: National Policy Committee.

Foro de Contadurias Gubernamentales de America Latina (FOCAL). 2007. *Canadian Perceptions of Latin America and the Caribbean: Survey Report October 2007.* Ottawa: FOCAL.

Gauthier, Chantal, and France Lord. 2013. *Engagées et solidaires : Les sœurs du Bon-Conseil à Cuba 1948–1998.* Montréal: Carte Blanche.

Gay, Daniel. 1983. *Les élites québécoises et l'Amérique latine.* Montréal: Nouvelle optique.

Gérin-Lajoie, Paul. 1976. *Plan of Action for Cooperation between Canada and Latin America: New Forms of Cooperation for Middle-Income Countries.* Ottawa: Government of Canada.

Graham, Bill. 2016. *The Call of the World: A Political Memoir.* Vancouver: University of British Columbia Press.

Granatstein, J.L. 1985. "The Anglocentrism of Canadian Diplomacy." In *Canadian Culture: International Dimensions,* edited by Andrew F. Cooper. Toronto: Canadian Institute of International Affairs.

Harbron, John. 1974. "Growing Pressures on Canada to Seek Hemispheric Identity." *International Perspectives,* May–June.

– 1976. "Canada Draws Closer to Latin America: A Cautious Involvement." In *Latin America's New Internationalism: The End of Hemisphere Isolation,* edited by Roger Fontaine and James Theberge. New York: Praeger.

Harris, Cole. 1966. "The Myth of the Land in Canadian Nationalism." In *Nationalism in Canada,* edited by Peter Russell. Toronto: McGraw-Hill.

Henrikson, Alan K. 1980. "The Geographical 'Mental Maps' of American Foreign Policy Makers." *International Political Science Review* 1.

Hillmer, Norman. 2015. *O.D. Skelton: A Portrait of Canadian Ambition.* Toronto: University of Toronto Press.

Holmes, John W. 1962. "Our Other Hemisphere: Reflections on the Bahia Conference." *International Journal* 17.

– 1968. "Canada and Pan America," *Journal of Inter-American Studies* 10.

– 1979. *The Shaping of Peace: Canada and the Search for World Order, 1943–1957,* vol. 1. Toronto: University of Toronto Press.

Irwin, W. Arthur Irwin. 1964. "Memorandum on Possible Entry of Canada into the Organization of American States." 17 July, LAC, RG 25, file 20–4-OAS-4–1 pt. 2.

– 1965. "Should Canada Join the Organization of American States?" *Queen's Quarterly* 72.

Kirk, John, and Peter McKenna. 1997. *Canada-Cuba Relations: The Other Good Neighbor Policy*. Gainesville: University Press of Florida.

Knowles, A.J. 1960. "The Poker-Faces of Latin America." *Saturday Night*, 1 October.

Latin American Working Group (LAWG). 1973. *Chile versus the Corporations*. Toronto: LAWG.

Léger, Jean-Marc. 1961. "Le Canada français et l'Amérique latine." *Le Devoir* 18 November.

LeGrand, Catherine. 2009. "L'axe missionnaire catholique entre le Québec et l'Amérique latine. Une exploration préliminaire." *Globe. Revue international d'études québécoises* 12.

– 2013. "Les réseaux missionaries et l'action sociale des Québécois en Amérique latine, 1945–1980." *Études d'histoire religieuse* 79.

Lubbock, Michael. 1976. "Canada and Latin America." *North-South* 1.

Lumsden, C. Ian. 1968. "The 'Free World' of Canada and Latin America." In *An Independent Foreign Policy for Canada?*, edited by Stephen Clarkson. Toronto: McClelland & Stewart.

Macdonald, Laura. 2016. "Canada in the Posthegemonic Hemisphere: Evaluating the Harper Government's Americas Strategy." *Studies in Political Economy* 97.

MacKay, R.A., and E.B. Rogers. 1938. *Canada Looks Abroad*. Toronto: Oxford University Press.

Martin, Paul. 1936. "Canada and the Collective System." In *Canada: The Empire and the League*. Toronto: Thomas Nelson & Sons.

Massey, Vincent. 1947. "Should Canada Join the Pan-American Union?" *Maclean's Magazine*, 15 August.

– 1948. "Canada and the Inter-American System." *Foreign Affairs* 26.

McFarlane, Peter. 1989. *Northern Shadows: Canadians and Central America*. Toronto: Between the Lines.

McKenna, Peter. 1995. *Canada and the OAS: From Dilettante to Full Partner*. Ottawa: Carleton University Press.

McKercher, Asa. 2012. "Southern Exposure: Diefenbaker, Latin America, and the Organization of American States." *Canadian Historical Review* 93.

– 2014. "'Ultimate Destiny' Delayed: The Liberals, the Organization of American States, and Canadian Foreign Policy, 1963–1968." *Diplomacy & Statecraft* 25.

– 2018. "A Limited Engagement: Diefenbaker, Canada, and Latin America's Cold War, 1957–63." In *Reassessing the Rogue Tory: Canadian Foreign Relations in the Diefenbaker Era*, edited by Janice Cavell and Ryan Touhey. Vancouver: University of British Columbia Press.

Miller, Eugene. 1947/8. "Canada and the Pan American Union." *International Journal* 3.

Ministerial Report. 1968. "Some Impressions of the Ministerial Mission to Latin America." LAC, Mitchell Sharp Papers, vol. 32, file Trips: Latin America Ministerial Mission.

Mongeau, Serge. 1974. "Lecons a retenir au Quebec." *Chili-Quebec Informations*.

Montreal Star. 1961. "Commentator Urges Early Membership in OAS." 13 December.

Mulroney, Brian. 2007. *Memoirs*. Toronto: McClelland & Stewart.

Murray, David. 1974. "Canada's First Diplomatic Missions in Latin America." *Journal of Interamerican Studies and World Affairs* 16.

Nef, Jorge, ed. 1978. *Canada and the Latin American Challenge*. Guelph: Ontario Co-operative Programme in Latin American Studies.

North, Liisa. 1990. *Between War and Peace in Central America*. Toronto: Between the Lines.

Nossal, Kim Richard. 1992. "A European Nation? The Life and Times of Atlanticism in Canada." In *Making a Difference: Canada's Foreign Policy in a Changing World Order*, edited by John English and Norman Hillmer. Toronto: Lester.

Ogelsby, J.C.M. 1969. "Canada and the Pan American Union: Twenty Years On." *International Journal* 24.

– 1976. *Gringos of the Far North*. Toronto: Macmillan.

– 1979. "A Trudeau Decade: Canadian-Latin American Relations, 1968–1978." *Journal of Interamerican Studies and World Affairs* 21.

Pearson, Lester. 1948–9. "Canada and the North Atlantic Alliance." *Foreign Affairs* 28.

– 1965. Press Conference in Jamaica, 30 November, The National Archives of the United Kingdom, FO 371/179159.

Peddie, Francis. 2014. *Young, Well-Educated and Adaptable: Chilean Exiles in Ontario and Quebec, 1973–2010*. Winnipeg: University of Manitoba Press.

Podea, Iris. 1948. "Pan-American Sentiment in French Canada." *International Journal* 3.

del Pozo, José. 2009. *Les Chiliens au Québec : Immigrants et réfugiés, de 1955 à nos jours*. Montréal: Boréal.

Protocol Division to Cadieux. 1966. "Call on Prime Minister by new Argentine Ambassador." 24 January, LAC, RG 25, file 20–4-OAS-4–1-pt. 4.

Report. 1948. "'Western Union' and an Atlantic Pact: A Survey of Recent Developments." 11 September, LAC, RG 2, file U-10–11.

Resnick, Philip. 1970. "The New Left in Ontario." In *The New Left in Canada*. Montreal: Black Rose Books.

Robertson to Diefenbaker. 1961. "Canadian and United States Relations with Latin America." 16 February, LAC, Basil Robinson Papers, vol. 4 file Feb 1961.

Rochlin, James. 2012. "Introduction – Canada and the Americas: There's still Much to Discover." In *Canada Looks South: In Search of an Americas Policy*, edited by Peter McKenna. Toronto: University of Toronto Press.

Roussin, Marcel. 1959. *Le Canada et le Système interaméricain*. Ottawa: Éditions de l'Université d'Ottawa.

– 1964. "Le Canada et l'OEA." *Cité Libre* 15.

Ryan, Claude. 1964. "Le Canada et l'Amérique latine." *Le Devoir*, 28 August.

Sclanders, Ian. 1960. "Canada Finally Joins the Western Hemisphere." *Maclean's*, 10 September.

Scott, Frank. 1938. *Canada Today*. Oxford: Oxford University Press.

– 1942. "Canada's Role in World Affairs." *Food for Thought* 2.

Scott, Jonathan. 2011. *When the Waves Ruled Britannia: Geography and Political Identities, 1500–1800*. Cambridge: Cambridge University Press.

Sheinin, David. 2009. "Cuba's Long Shadow: The Progressive Church Movement and Canadian-Latin American Relations, 1970–1987." In *Our Place in the Sun: Canada and Cuba in the Castro Era*, edited by Robert Wright and Lana Wylie. Toronto: University of Toronto Press.

Smith, David Edward. 1966. "Should Canada Join the Organization of American States?" *Queen's Quarterly* 73.

Soderlund, Walter. 1976. "Latin American Studies in Canada: Ten Years after the Initial Assessment." *North-South* 1.

Soward, F.H., and A.M. Macaulay. 1948. *Canada and the Pan American System*. Toronto: Ryerson Press.

Stevenson, Brian. 2000. *Canada, Latin America, and the New Internationalism: A Foreign Policy Analysis, 1968–1999*. Montreal: McGill-Queen's University Press.

Swift, Jamie. 1977. *The Big Nickel: Inco at Home and Abroad*. Toronto: Between the Lines.

Taber, Jane. 2011. "Harper Spins a New Brand of Patriotism." *Globe and Mail*, 19 August. Accessed 20 October 2021. https://www.theglobeandmail.com /news/politics/ottawa-notebook/harper-spins-a-new-brand-of-patriotism /article618385/.

Thérien, Jean-Philippe, and Gordon Mace. 2013. "Identity and Foreign Policy: Canada as a Nation of the Americas." *Latin American Politics and Society* 55.

Vallières, Pierre. 1967. "Cuba révolutionnaire." *Parti Pris*, September.

Vucetic, Srdjan. 2011. *The Anglosphere: A Genealogy of a Racialized Identity in International Relations*. Palo Alto: Stanford University Press.

Webster, David. 2016. "Foreign Policy, Diplomacy, and Decolonization." In *Canada and the Third World: Overlapping Histories*, edited by Karen Dubinsky, Sean Mills, and Scott Rutherford. Toronto: University of Toronto Press.

Wilgress, Dana. 1953. "The Organization of American States." 26 February, LAC, RG 25, file 2226–40.

– 1967. *Memoirs*. Toronto: Ryerson Press.

Wright, Cynthia. 2009. "Between Nation and Empire: The Fair Play for Cuba Committees and the Making of Canada-Cuba Solidarity in the 1960s." In *Our Place in the Sun: Canada and Cuba in the Castro Era*, edited by Robert Wright and Lana Wylie, 96–120. Toronto: University of Toronto Press.

3 Life-Death-Rebirth: The Latin American Working Group and Civil Society Relations with Latin America[1]

JOHN W. FOSTER

Canada's relations with the Americas have been shaped not just by government and corporate actors but also by civil society actors who worked to support civil society counterparts in the Americas and to lobby for changes in Canadian government policies. One example is the thirty-year work of the Latin American Working Group (LAWG), a research and solidarity organization formed in 1966. Created in response to appeals of counterparts affected by the 1965 US invasion of the Dominican Republic, the group aimed to discover Canada's role in hemispheric events, educate Canadians on the socio-economic realities of the people of Latin America, and to support their struggles for social justice.

LAWG focused its research on Canada's role in the region, with special emphasis on Canadian aid policies and the role of multinational corporations. LAWG worked in solidarity with the Canadian churches, trade unions, and a wide variety of other non-governmental and grass roots organizations to raise public awareness and to lobby for changes in Canadian government policies. The group remained independent of government funding, benefitting from modest contributions from the United and other churches, some religious orders and project funding where, for example, a tour to the region involved Canadian trade unionists.

This chapter reviews the origins and evolution of the LAWG collective, based on original archival research from the LAWG organizational archive held by the Clara Thomas Archives and Special Collections at York University and the LAWG library collection at CERLAC (York University). Particularly useful are the materials available at the LAWG history project website (https://lawghistoryproject.wixsite.com/law ghaycaminohistory), including interviews with former staff, original activists, and observers. A summary history of the organization is available (Acton and Anderson 2017), as is a graphic PowerPoint presentation on the origins and formation of the group (Foster 2015).

These sources provide windows onto the evolution of Canadian civil society interest in Latin America and onto the nature and careers of solidarity organizations, but they also contain a vast body of documents and publications produced by Latin American organizations and by human rights, labour, and solidarity bodies involved in the links between Canadians and Latin American counterparts.

This chapter examines the origins of the group, situated in the explosion of youth activism associated with the "sixties," with roots in Christian youth movements. It explores similarities and differences with the coincidental emergence of NACLA in the United States. It also notes the responses of Canadian volunteers to epic external events, including the 1973 military coup in Chile and the revolutionary events in Central America in the 1970s and 1980s. In addition to the provision of information in a pre-internet era, LAWG was a source of political assessment and advice not only for other solidarity groups but for major Canadian religious denominations, NGOs (non-governmental organizations), and ultimately significant trade union sectors. The role of the group in organizing and/or advising fact-finding and educational tours by scores of Canadians contributed to the expansion of a knowledgeable and committed constituency across Canada. Building on study and research, LAWG played an instrumental role in challenging the Canadian government's response to the military coup in Chile, particularly contributions to a change in policy which resulted in thousands of refugees and a number of political prisoners coming to Canada.

To cite one sector of the Canadian constituency, the parties in the ecumenical Ten Days for World Development through several years of its focus on Central America, developed policy proposals for Canada with regard to the defence of human rights, refugee assistance, intiatives for peace, and development aid. These were carried forward in direct representations to the government, engagement with politicians and the press, and hosting educational visits by Central Americans.

The author is a participant/observer, who was national President of KAIROS at the moment of conception, was an active volunteer as a graduate student, then a professional employee in the United Church and instrumental in the leadership of inter-church coalitions addressing Canadian foreign, human rights, and refugee policy. These roles included repeated policy advocacy vis-à-vis the Canadian government as well as advocacy in the annual Geneva sessions of the UN Human Rights Commission. More recently he was an active member of the *Sí*,

Hay Camino history project through 2018. With LAWG, he was always a volunteer and collective member, never a staff person but often a "bridge" figure linking LAWG to ecumenical coalitions, to initiatives like Missions for Peace and the Roundtable on Negotiations for Peace in Central America (Baranyi and Foster 2012), as well as participating in periodic hemispheric and trinational civil society events as well as academic conferences on Latin America. [2]

Approach

At the inauguration (16 February 2017) of the LAWG library documentary collection at the Centre for Research on Latin America and the Caribbean (CERLAC) at York University, Prof. Luis van Isschot, a former LAWG volunteer, quoted the Haitian scholar Michel-Rolph Trouillot on the creation of history and how silences and forgetting are transcended: "Silences enter the process of historical production at four crucial moments: the moment of fact creation (the making of sources); the moment of fact assembly (the making of archives); the moment of fact retrieval (the making of narratives); and the moment of retrospective significance (the making of history in the final instance)" (Trouillot 1995, 26). This chapter establishes the outline of a *narrative* based on selective retrieval of facts derived from the sources created by the LAWG during a thirty-year history.

The chapter concludes with reference to the making of *archives* by employees, members, and alumni and outlines a history of the enterprise, inviting others to investigate from diverse points of view. The author also reflects on the "retrospective significance" of the work of the collective.

Historical Context

Ferment at home

The origins of LAWG have their roots in the social and political ferment of the early 1960s in Canada and around the world. This period was characterized by a certain "vogue" for youth, breaking away from conventional aspects of the previous decade (Rowntree 1968). International events, like the Cuban revolution and the deepening conflicts in Vietnam, together with the struggle for civil rights in the United States and the March on Washington, were part of a heady mix. Students in the United States organized Students for a Democratic Society (SDS), issuing a declaration – the *Port Huron Statement* – in 1962. The Student Non-Violent Coordinating Committee (SNCC), formed in 1960 in the wake of the Christian Leadership Conference, was followed by a Canadian support group, "Friends of SNCC." The Fellowship of Reconciliation (FOR) brought together a group of Canadian and US youth and student leaders with "mentors" like Bayard Rustin and A.J. Muste in 1963, in which a number of future leaders, like Tom Hayden, were present. In 1964 a delegation of two dozen Canadian youth leaders from the peace movement and student and church organizations took part in the World Assembly of Youth (University of Massachusetts, Amherst) listening to Martin Luther King Jr. and the Freedom Singers and protesting US bombing of North Viet Nam.

The ferment spreading among youth and spilling across the border is symbolized by the metamorphosis in several key Canadian organizations: The National Federation of Canadian University Students (NFCUS) became the Canadian Union of Students with a more activist agenda. The Combined University Campaign for Nuclear Disarmament (CUCND) became the Student Union for Peace Action (SUPA). The United Church Young People's Union (YPU) became KAIROS. In December 1965 members of SUPA, the Student Christian Movement, and KAIROS met simultaneously in Saskatoon with some common and some separate sessions. In May 1966 the *Toronto Star's Canadian Weekly* supplement featured profiles of a number of youth leaders (including two from KAIROS and one from CUS), with the title "What's bugging the committed kids?" After promising an initiative in 1965, in 1966 the federal government announced a youth volunteer programme designed to channel the demands for "change": the Company of Young Canadians (CYC) (Hamilton 1970). Ideological currents affecting many, if not all, in these organizations were increasingly identified as "New Left," responding to a call from the formative days of the Students for a Democratic Society (Kostash 1980). In summary, these organizations brought together concerns about domestic and international social justice, including a critique of US imperialism in the Western Hemisphere and elsewhere. A popular examination of Canada's subservience to US global power was central to the 1960 study *Peacemaker or Powdermonkey* by CBC journalist James M. Minifie (Moffatt n.d.). For those developing LAWG's orientation, a key work grew out of an encounter at Union Theological Seminary in New York City, and a commission by the University Christian Movement. Student leader Carl Ogelsby of SDS and theologian Richard Shaull of Princeton University published *Containment and Change* (1967), which examined the Cold War, Vietnam, imperialism, and social revolution.

Impulses from Abroad

Events in Latin America, not only the Cuban Revolution, Bay of Pigs, and missile crisis, but also the military coup d'état in Brazil (1964), and the US Marine invasion and occupation of the Dominican Republic (1965) provoked responses among youth and students. The American counterpart of the Canadian Student Christian Movement – the University Christian Movement (UCM) – issued a call for a Task Force on US–Latin American relations, to undertake research and study, public education, and lobbying. Both the UCM and the SCM sponsored study tours to Cuba in the early 1960s (Douville 2011; Kostash 1980). An American employed by both the United States and the Canadian Fellowship of Reconciliation, Brewster Kneen, spent the early 1960s visiting campuses and listening for trends in student interests. Married to a Canadian, he visited the 1965 summer conference of the Student Union for Peace Action (SUPA) in Quebec, along with some young veterans of civil rights actions in the United States. In the months following, while organizing an event related to Cuba, the Kneens began to convene a small group for Bible study, Judeo-Christian observances, and political discussion in Toronto called the Christian Left. Several members were leaders in CUS, KAIROS, and other youth organizations, including several who became part of the LAWG.

Making Links

Early in 1966 the United Church KAIROS received an invitation from the regional Protestant network ULAJE (Union of Latin American Evangelical Youth) to participate in its pan-hemispheric conference in Puerto Rico later that year. ULAJE's staff had worked to build bridges between youth in revolutionary Cuba and the United States. The Canadians attending the Puerto Rico congress were challenged and befriended by counterparts from the Dominican Republic and came back to Canada with ideas of common study and "work camp"–style projects to propose. When attending a KAIROS national conference "Summer Event," they got support for the creation of a working group on international affairs that would, among other things, follow up with ULAJE and the Dominican Republic counterparts, develop a project with Cuba, and prepare educational materials. This working group, focusing more clearly on the hemisphere, gradually became known as LAWG, the Latin American Working Group.

Coincident to this initiative, two American students took part in election observation in the Dominican Republic in 1966, convinced of the need for independent and critical information and analysis. On their return, Fred Goff, a son of Protestant missionaries, toured US campuses in search of allies. Picking up on the UCM Latin American Concerns Committee, Princeton theologian Richard Shaull (who had also been in the Dominican Republic during the election) hosted a two-day conference at his university in October. Shortly thereafter (5–6 November) a gathering of 23 people (including one Canadian from LAWG) established what became the North American Congress on Latin America (NACLA), which survives today.

A couple of observations on the characteristics of these developments are in order: Both were led by youth and had university and academic connections; both had close connections with Protestant church bodies and the SCM/UCM but were not restricted to Christian believers. Both were concerned with critical analysis, education, and publications. These young people tended to be white, both student and non-student, male and female. They had limited experience with the fairly small Latin American diaspora in Canada but were engaged with Latin American graduate students in Toronto and the *Prensa Latina* outpost in Montreal.

NACLA founder and staff member Fred Goff came to Canada to lead a Toronto workshop dealing with the Caribbean, Cuba, and the Dominican Republic. NACLA Chair Richard Shaull was a resource person for the LAWG-sponsored workshop at Westport, Ontario, in May 1967. As one NACLA founder[3] commented, it brought together "Christers" and secular folk moved by analyses of power structures dominating North American societies, like sociologist C. Wright Mills, author of *The Power Elite*, who launched a critique of imperialism with *Listen Yankee* in 1960 (Mills 1956; 1960). In his history of NACLA, Fred Rosen summarizes the orientation of the time: "Looking toward Latin America (for many North Americans, discovering Latin America for the first time), the student movement developed strong sympathies for groups seeking to liberate themselves from dictatorship, underdevelopment and 'Yanqui' imperialism." NACLA moved swiftly from an initial focus on the Dominican Republic and US foreign policy to "focus on the structure of Latin American political institutions ... and from there to a critique of what NACLA came to call 'the Empire'" (Rosen 2002, 15).

Following on the orientation embodied in *Containment and Change*, Canadian Bruce Douville in his doctoral dissertation commented that despite the aura of the era, "the religious roots of the New Left were not Zen, eastern mysticism or the occult, but rather the Social Gospel, Christian existentialism, and social Catholicism" (2011, 380ff). NACLA and LAWG continued on parallel but not identical tracks for the next three decades.

1965–1972 Volunteer Activism

Initially LAWG was made up of volunteers from several eastern Canadian cities and with some limited church support was able to employ a part-time staff person. Toronto-based volunteers began to meet weekly in people's homes with periodic workshops and retreats with allies in other centres: "LAWG began to explore what solidarity means, and how to go beyond humanitarian support" (Acton and Anderson 2017, 4). Some volunteers participated in overseas work/study projects by organizations like the Conference of Inter-American Student Projects (http://www.ciasp.ca/), a largely Catholic body. However, very shortly they became critical of what was referred to as "assistantialism," traditional charity work. Several associates developed deeper connections and personal relationships in the Dominican Republic, some of these cultivated through missionary appointments in the south. Meanwhile internal study ventured into "works of Gramsci, Gunder Frank and other theorists developing a deeper economic analysis and understanding of how imperialism perpetuates underdevelopment" (Acton and Anderson 2017, 5).

Of particular importance was the engagement of several LAWG members in developing a practice of "conjunctural analysis" of Canadian and international affairs. Modelled on the Mexican non-governmental organization (NGO) CENCOS (*Centro Nacional de Comunicación Social, AC*), which also had religious links, the Canadian News Synthesis Project (CNSP) analysed a range of Canadian publications and published a monthly review based on a strict thematic structure.

LAWG was increasingly devoted to research, not only on Latin America but on Canadian roles in the hemisphere, including foreign aid, investment, and the activities of Canadian corporations, including Falconbridge, INCO, and Brascan. Publications included *Chile vs the Corporations* (co-sponsored by the Development Education Centre, DEC), *The Brascan File* (Project Brazil, 1973), a book-length study *Falconbridge*, as well as *LAWG Inside Report*, the *LAWG Labour Report*, and the *LAWG Newsletter*. In concert with the Toronto-based Women's Press, LAWG collaborated on three books: *Cuban Women Now* (Randall 1974), *Population Target* (Mass 1976), and *Central American Women Speak for Themselves* (LAWG 1983).

In 1972 the Missions Office of the Canadian Catholic Bishops Conference held an event – "Solidarity 72" – in Mississauga, involving key Latin American partners and Canadian allies from lay and youth associations in Canada. The event proved to be a launching platform for collaboration between the Protestant and secular LAWG member-ship and several Catholic youth, members of religious and missionary orders, and key leadership from the Bishops Conference and the Scar-boro Foreign Mission Society.

The intellectual and theological analysis of the group was reinforced by an event at the University of Toronto titled "Imperialism and Change in Latin America," involving Canadian and visiting Latin Ameri-can academics from York University and the University of Toronto, among others; and by the input of theology of liberation there and at the Solidarity 72 event. LAWG began plans to organize a tour to Chile, then governed by democratically elected socialist president Salvador Allende, and two LAWG members visited the UNCTAD Conference in Santiago in 1972. Chile became of special interest to LAWG as a result of the participation of Chilean Methodist Arturo Chacón and his partner Canadian Florrie Snow during their studies at University of Toronto in the early 1970s.

1973–1974 The Cauldron: Coup d'état and Aftermath

The period of LAWG's formation, which began in the mid-sixties, was interrupted and transformed by the 11 September 1973 *coup d'état* in Chile. The coup marked a period of deep transition in the life of LAWG, the *first* extensive reorientation of the focus and make-up of the group. A *second* period of change marked the later 1970s, and a *third* in the late 1980s, leading ultimately to dissolution and new formations.

The immediate response to the coup of 11 September brought together students (both Canadian and international), academics, church folk, and activists of various political orientations in direct lobbying, political pressure, and soon refugee reception. The Inter-Church ad hoc Com-mittee on Chile took shape among denominational officials and the Canadian Council of Churches, as did the Toronto Welcome Commit-tee for Refugees from Chile, based at the Centre for Spanish-Speaking Peoples. The author was a founding member of the former and volun-teer with the latter. Both were instrumental in representations to the federal government. LAWG volunteers were active in both and pro-vided services with refugees from the landing of the first plane from the south, including simple essentials for winter life in Canada, accom-paniment in dealing with officials and doctors and accommodation and language training. Some LAWG members decided just to focus on this work, and several deep relationships and several binational marriages developed. Some refugees, deeply involved in secular left politics in Chile, found the involvement of many Christians in the refugee reception activities surprising. But the unique contribution of LAWG lay in communication.[4]

With former LAWG members in Santiago, LAWG staff member Tim Draimin made the links between the eyewitness testimony of the Chacóns at street level in Santiago and the Canadian media, in particular CBC's *As it Happens*. Eyewitness testimony drove public pressure against the Trudeau government's refusal to initiate a major programme for refugees. LAWG volunteers were intensely involved in public education events and then in refugee reception and settlement.

Reflective reviews on this period were presented at the 40th Anniversary Remembrance of the Casa Salvador Allende in Toronto (7 September 2013), both by figures active in the churches and in LAWG. They included details of the period of intense advocacy leading to the approval of a major movement of refugees and the development of a unique programme for the selection and transfer of several hundred political prisoners and their families to Canada (Gunn 2018; see also speeches from the Toronto anniversary event by George Cram and John Foster, accessible at https://lawghistoryproject.wixsite.com /lawghaycaminohistory).

With the arrival of several thousand refugees, demands on volunteers for help with fundamental issues of health, housing, language, etc. were extensive. Further complications arose as a number of refugees organized partisan formations in exile, embodying member parties of the Popular Unity coalition in Chile, including sub-party tendencies and factions. Some volunteers made the transition away from refugee support and back to research, writing, and political advocacy. Acton and Anderson (2017, 6) state, "LAWG took the decision to maintain a non-sectarian and unitarian stance. The decision ... was not an easy one, nor fully resolved, as a result several members left LAWG." When, late in the 1970s, a LAWG member (the author) decided to run for Parliament for the NDP, tensions were "further exacerbated," and "in keeping with its position to remain non-aligned, LAWG chose not to support the candidate as an organization, only as individuals" (Acton and Anderson 2017, 6).

Alongside this political clarification, LAWG "emerged with greater clarity about solidarity as a relationship of mutuality and 'partnership' ... increasingly seen as the building of links between social movements in Canada and the South, with the recognition that in both Canada and Latin America 'we each have our own historical projects'" (Acton and Anderson 2017, 6).

The question of *how* to carry forward this solidarity developed into what was termed a "sectoral" approach. The early connections with the churches and ecumenical coalitions were followed by the formation of links with workers and worker organizations struggling with the same transnational corporation. LAWG members had established relations with workers and communities where Falconbridge was established in the Dominican Republic, undertook a detailed corporate research project on the company at home and abroad, and established relations with its Canadian workers in the Mine, Mill and Smelter Union in Sudbury.

Given the intensity of the engagements and the political cross-currents of the time, the work of LAWG volunteers and staff took the form of a *collective* which could be joined only after an interview/approval process and commitment to the non-sectarian orientation. Weekly meetings and study sessions (usually on Monday evenings) brought together ten or more members for several decades. The coup and post-coup experience transformed the organization into a more professional team and tighter formation overall (Acton and Anderson 2017, 5). In his study of Canadians and Central America, Peter McFarlane concludes: "By the early 1970s LAWG included a core of researchers, writers, and academics who would become Canada's foremost experts on Central and South American affairs" (1989, 141). To maintain critical distance from official policy, LAWG refused to seek/accept funding from governments but continued to receive support from denominational churches and religious orders.

LAWG played an important role bringing together an alliance of religious, NGO and other activists. McFarlane notes that "the Chilean issue helped to weld these distinct organizations into a Latin American lobby in Canada" (1989, 142). During this period research, advocacy, orientation of fact-finders and transnational linkages with southern counterparts continued not only with Chile but in response to military coups in Argentina and Uruguay as well as Brazil. The church human rights initiative was also moved by these tragic developments and in 1976 sponsored a tri-party group of MPs on a mission south. This resulted in a significant report, "One Gigantic Prison" (Brewin 1976), and the reformation of the Inter-Church Committee on Chile into the Inter-Church Committee on Human Rights in Latin America (ICCHRLA), involving several national denominations along with both male and female religious orders (Simalchik 1993). While much of the advocacy of these organizations was directed at the Canadian federal government, human rights findings which resulted from periodic fact-finding missions and relations with sister organizations in Latin America were taken forward in detailed representations to the Human Rights Commission of the United Nations.

Refocus: 1979–1989 Central America

Events in the Southern cone continued to preoccupy LAWG members in the later 1970s, with particular contributions on their analysis of conditions in the Chilean economy and society under Pinochet. However, the humanitarian and political crises in Central America were increasingly known, as was the often central role of religious actors and the articulation of liberation theology, documented in such works as Penny Lernoux's *Cry of the People* (Lernoux 1980). Political processes developing in the late 1970s, particularly the progress of the Sandinista revolutionaries against Somoza in Nicaragua and the Frente Sandinista de Liberación Nacional (FSLN) victory in 1979 motivated growing activism across Canada in solidarity with these struggles. The growth of popular movements in El Salvador and Guatemala, the situation of refugees in the region, the escalation of repression in both countries, and the role of the United States along with the relative passivity of the Canadian government engaged LAWG, and many allies, in fact-finding, identification of allies and links, public education and political advocacy aimed at gaining recognition of the legitimacy of movements of resistance in the region, and accompanying representatives of opposition forces in El Salvador and Guatemala in visits to Canadian officials and the Canadian media.

Many Faces

As LAWG collective members and former staff Acton and Anderson point out, fundamental to LAWG's credibility and usefulness were several roles the group was able to play:

Information and orientation:

LAWG maintained a resource centre and library as well as a continuing and varied programme of research, and contributed to public education about the hemisphere through the organizing of numerous tours of Canadians to Central America and Mexico, and visits from political, union and civil society figures from the south to Canada. LAWG contributed to orientation for fact-finding missions of other organizations, as well as detailed information for refugee lawyers and tribunals. A close working relationship with a number of development NGOs was a central part of the growth of the multi-faceted Central American movement in Canada, including collaboration with Tools for Peace across the country.

(Acton and Anderson 2017, 7)

The cumulative impact of this work was considerable. As a key participant/observer of the decade following the FSLN victory, Jesuit, now Cardinal, Michael Czerny assessed: "Canadians by the tens, then hundreds and now thousands – students, academics, parliamentarians, religious and labour leaders, interested citizens – have been visiting Central America and have been welcoming Central American visitors and refugees to Canada. New awareness and mutual respect were generated, as were new expectations" (North 1990, 260). Brian Stevenson notes that the extent of citizen response to the Chilean coup d'état and the growing engagement with Central America represented an "assault" on Canadian foreign policy and a continued critique of Canadian corporations' operations abroad (Stevenson 2000, 196). John Graham, the key "point person" at Foreign Affairs remarks on the surprising extent of public concern: "never before or since has Canadian civil society become so exercised about a foreign policy issue. A parliamentary committee, mandated to sound public perceptions ... was astonished to find that Central America topped Canadian concerns above apartheid in South Africa, nuclear disarmament, the Soviet Union and Eastern Europe" (Graham 2015, 127).

In 1981 the newly formed Canada Latin America Resource Centre (CLARC) was established as an independent entity facilitating fund-seeking from sources that LAWG itself was not approaching, sharing the new LAWG office on Toronto's Harbord Street. In this pre-internet era, CLARC engaged a librarian to manage a growing collection of hard-copy newsletters, reports, pamphlets, and books which was a "real draw for researchers, media, students and church people." LAWG/CLARC supported a bookstore service (in person and by mail), the resource centre files as well as publications and a library (Acton and Anderson 2017, 7).

Another important source of impact for LAWG was its engagement with Canadian mainline churches. Given its origins and the initial dependence on denominational budgetary support, however limited, LAWG was well-positioned to engage with the generation of new inter-denominational formations which characterized the 1970s. In 1972 an Inter-Church Committee on Development Education met to consider utilizing the Lenten period for focused study and events. This initiative became known as Ten Days for World Development and commenced with a cross-Canada tour of church leaders from 8 March 1973

(Mihevc and Lynd 1994; Gunn 2018). Of ten ecumenical bodies formed in the period, four were of particular relevance: the Inter-Church Committee on Chile (ICCC) (later the Inter-Church Committee on Human Rights in Latin America [ICCHRLA]), the Task Force on the Churches and Corporate Responsibility (TCCR), Ten Days for World Development, and GATT-Fly (later the Ecumenical Coalition for Economic Justice, ECEJ).[5] In some cases LAWG was a member at the table, in others it provided volunteer or contracted research, including detailed economic analysis of the Chilean dictatorship. LAWG personnel were actively involved with the formation of the ICCC and ICCHRLA and from time to time assisted with orienting both Canadian and Latin American visitors. In the case of the several years in which Ten Days focused on Central America, LAWG assisted with strategic planning and research, writing, and public speaking. The Ten Days network in dozens of Canadian communities exercised remarkable pressure on the positions and actions of the Canadian government in the years in which it focused on Central America. During the 1980s campaigns were undertaken in defence of human rights in the region, in opposition to US support for repressive governments in the region, and to encourage pressure on the Canadian government to support strong action by the United Nations Human Rights Commission as well as in its bilateral relationships.

Policy advocacy, leadership, and collaboration:

Whether initiated by LAWG or engaging LAWG as a coalitional ally, the late 1970s and 1980s were scenes of new efforts devoted to peace in Central America, diplomatic and economic policy alternatives, and citizen activism. LAWG members and allies in other NGOs and agencies like the Jesuit Centre worked in concert in a number of these initiatives:

- *Chile-Canada Solidarity* – a campaign arm open to volunteers across Canada;
- *Central America Update* – a bulletin published in partnership with the Jesuit Centre, a key distribution means for front-line news and analysis;
- *NICA* –Non-Intervention in Central America – a campaign arm to enlist public pressure;
- *Mission for Peace* – highlighting the need for Canadian diplomatic and public action for peace in the Central American conflicts, utilizing fact-finding tours by Canadian opinion leaders;
- *CAPA – Canada–Caribbean–Central America Policy Alternatives* – collaboration with academics and NGO leadership in research

and advocacy to promote alternatives to the US intervention in El Salvador, Nicaragua, and the region (Baranyi and Foster 2012);

- *The Roundtable on Negotiations for Peace in Central America* – involving Canada–Caribbean–Central American Policy Alternatives (CAPA), an alliance with academics, supported by public funds, seeking allies in the US and Latin America to deepen policy advice and initiatives. With support from the Canadian Institute for International Peace and Security, CAPA held a series of invitational roundtables in Ottawa between 1985 and 1990 publishing a series of policy studies and a book (North 1990). The initial objective was "to discuss possible policy initiatives for bilateral and multilateral action, likely to de-escalate conflicts in Central America and to strengthen the Contadora peace process" (North 1985).

NETWORK SUPPORT

The rise of the right in Thatcher's United Kingdom and Ronald Reagan's United States put solidarity groups and often development NGOs on the defensive. Following the FSLN revolution in 1979, LAWG's engagement with popular movements in Central America deepened, as did its response to escalating repression in Guatemala and El Salvador (Acton and Anderson 2017). The continued role of information, "intelligence" and analysis was central to the contribution that LAWG members sought to animate. LAWG members collaborated with Canadian NGO representatives and counterparts in the region (including CUSO, OXFAM, and Development and Peace) and allies across Canada: "LAWG had access to information and contacts in Central America that could strategically feed into the solidarity work of groups in the Maritimes, the Prairie provinces, Quebec and BC" (Acton and Anderson 2017, 8). In the latter case LAWG assisted in facilitating the first tour by BCNic to Nicaragua in 1981, "providing an opportunity for some of BC's activists and labour leaders to see for themselves what was happening with people's struggles in Central America, and offering a model for the type of sectoral work that might be employed elsewhere in Canada" (Acton and Anderson 2017, 8). Collaboration and support for a diversity of independent initiatives continued, including Medical Aid to Nicaragua (MATN), Tools for Peace and ICCSASW (International Commission for Coordination of Solidarity Among Sugar Workers).

The increased maturity of work among civil society, particularly among development NGOs and some solidarity organizations led to increased importance of coordination among them. In the mid-1980s LAWG was instrumental in organizing retreats which offered "an important mechanism for strategic discussions." "In these," Acton and Anderson (2017, 9) comment, "LAWG, while it had few resources, played a strategic convening role." Additional intelligence and analysis were provided through the establishment of an office and staff person in Mexico, supported by church, NGO, and LAWG resources (Acton and Anderson 2017, 9).

While detailed policy recommendations emerged in diverse fashion through policy briefs, fact-finding reports and denominational and union representations, fundamental principles could be identified running through most. Michael Czerny states that expectations emerging from these various sources "find an important echo and expression ... in fundamentals of foreign policy towards Central America." The basic tenets bear reiteration and include:

- Recognition of the indigenous socio-economic roots of the conflict;
- Respect for international law and its institutions, such as the International Court of Justice;
- Commitment to the peaceful resolution of conflicts;
- Insistence upon the principle of self-determination and non-intervention;
- Encouragement for international economic policies to promote equitable development and democratization; and
- Involvement in international efforts to bring peace to the region.

To these might be added respect for and implementation of fundamental human rights with protection of human rights defenders and the protection and safety of refugees (North 1990, 260).

LABOUR SOLIDARITY

While LAWG engagement with the churches, religious orders, and ecumenical formations continued, increased volunteer effort and staff time was devoted to labour, "developing a network of solidarity activists and contacts in most major unions, in cooperation with a national labour network" (Acton and Anderson 2017, 9). Activists confronted the position of the Canadian Labour Congress, which was less than supportive of Latin American unions who were not part of the Cold War era and US-dominated International Congress of Free Trade Unions regional formation in the hemisphere. There was also the legacy of US support for reactionary and anti-left union leadership in several instances. The challenge for LAWG and allies was to identify progressive elements and "moments" among workers' organizations in the south. "The challenge was to find new ways of educating and encouraging Canadian labour leaders and rank and file to provide solidarity to Central American workers. Solidarity tours … became a central strategy to change not only Canadian government policy, but also official labour policy" (Acton and Anderson 2017, 8).

Supported by a church-based fund in the United States, LAWG representative Louise Casselman used a Mexico base to explore the potential for deeper links between progressive sectors of labour movements in Central America and counterparts in Canada. This careful nourishing and informing of fact-finding and solidarity tours in both directions also resulted in development of union international policy at home.

Pivot: Mexico, New Alliances, New Focus

Mexico was of increasing importance in the 1980s, as a base for Central American solidarity and refugee reception, and the Mexican diplomats played an important role in supporting efforts to highlight human rights in the UN context as well as initiatives towards peace and against US intervention in the region. With frequent LAWG and solidarity tour visits in Mexico City, relationships were established, as well, with sympathetic Mexicans, including veterans of CENCOS news analysis, and figures like Adolfo Aguilar Zinser, who wrote critically about the situation and treatment of Guatemalan refugees in Mexico, human rights leaders like MariClaire Acosta, and academics interested in Canada, like María Teresa Gutiérrez Haces.

Simultaneously, Canadian society was gripped by debate over the Reagan–Mulroney bilateral trade proposal (CUSFTA), which engaged LAWG members,[6] as the likelihood of an extension to Mexico appeared as a possibility. The importance of the *maquiladoras* as sites for "runaway" transfer of manufacturing production and jobs began to preoccupy some in Canadian unions. The need for additional research was increasingly urgent.

In September 1988 John Dillon, a LAWG and ecumenical coalition ally, and two others undertook a fact-finding visit to Mexico, as part of a new LAWG-initiated project called Common Frontiers. The group visited the US/Mexican frontier to "investigate the implications of the Maquiladora free trade zones for Canada under the Canada–US FTA. The mission sought contacts among Mexican worker organizations and popular organizations "to explore ways to work together to improve wages, working conditions, living standards and environmental quality in both countries" (Common Frontiers 1989).

The tour returned with a report that was injected into the political debate in Canada. At the time, while the possibility of a North American

Free Trade Agreement (NAFTA) seemed on the horizon, many activists (mistakenly) thought it might be ten years off.

The legacy of this initiative was a project of research into economic and social impacts, actual and potential; publication of working papers on investment, policy alternatives, and other themes; and the provision of educational materials for the Pro-Canada Network and its various affiliates. The project included more fact-finding visits to Mexico, visits to Canada by Mexican analysts, including Adolfo Aguilar Zinser and José Luis Canchola, and the planning of a bilateral seminar in Mexico in September 1990. Mexican allies were established in the *Centro de Coordinación de Proyectos Ecuménicos* (CECOPE) and the CIEMs (Centres for Information and Migration Studies) in Tijuana, Ciudad Juárez, and Reynosa. A project for a Mexico City office, "*Fronteras Comunes Mexico-Canadá*" was launched (CECOPE 1989). A LAWG volunteer, Nick Kerestesi, took up temporary residence in Mexico to nurture this new work. The intention was to explore the opportunities to establish a coalition of organizations who might benefit from the experience of Canadian groups in challenging the bilateral "free trade" negotiations prior to 1988, and to develop joint monitoring, analysis, and critique of bilateral and trilateral negotiations.

While Common Frontiers staff resided initially in the basement of the LAWG office, utilized LAWG contacts and allies in Mexico, and proposals were circulated on LAWG letterhead, the initiative rapidly outgrew what LAWG could meaningfully support. Support did arise from the Canadian Auto Workers, the Canadian Centre for Policy Alternatives, CUSO, the Confederation of Canadian Unions, the National Action Committee on the Status of Women, the United Steelworkers of America, and many others. Working relations were established with Washington-based bodies, including the Institute for Policy Studies. In 1990 two key events were organized in Mexico City: a high-level binational seminar organized in part by forces related to the opposition PAN party and engaging Canadian political and non-governmental supporters and opponents of the FTA, and a binational *"Encuentro"* hosted at the headquarters of the FAT (*Frente Auténtico de Trabajo*) independent union, involving Canadians and Mexicans from a variety of sectors. Relationships were established, particularly at the second of these events, which continued through a series of bilateral and trilateral events over the coming years, which in a number of cases shadowed the official trade and investment negotiations.[7]

Looking back to the last century, there are emphases which continue, in some cases embodied in the same people. The clearest is the challenge to globalization embodied in the hemispheric trade and investment negotiations. Following on the trinational alliance challenging the NAFTA, links were established with Brazilian and other trade union and social movements, which found useful forums in the World Social Forum periodic gatherings. As had been the practice vis-à-vis NAFTA, with the coalitions and sectoral organizations negotiating common positions in critique of and representation against the NAFTA, so positions across much greater diversity were developed and represented to governments negotiating the proposed Free Trade Area of the Americas. A popular base for this opposition was on view in the streets of the Quebec City Summit of the Americas in April 2001. When the official negotiations failed, the civil society bodies could assume some of the credit. The civil society emphasis on human rights focused in part on Central America in the 1990s, continued in dialogue with both the Harper and Trudeau governments but with Colombia often the country of focus, emphasis on corporate social responsibility on the part of Canadian firms, and the development of a matrix for measuring the impact on human rights of investment in such fields as mining and trade and investment agreements themselves.

An independent project, the Toronto-based Maquila Solidarity Network, was also established in the early 1990s, commencing decades-long advocacy and support for Mexican workers and their organizations.

1990–1997 LAWG's Final Years

As a new decade began, LAWG members began a critical review of their future. A national and international consultation with partners was undertaken in 1991–2. Factors which influenced the review, directly and indirectly included:

- Collective members were often employed in high-pressure agency or union positions, at home and in the South, and
- The nature of informational work was changing with the emergence of the internet and email.

The transition was made from a collective meeting weekly to an open membership organization, meeting monthly, with a small executive as well as an annual meeting at which members outside of Toronto were able to attend. The resource centre was allied with the parallel work of the Toronto Committee for the Liberation of Southern African Colonies (TCLSAC) in the formation of the Counterpoint Resource Centre. A new building was purchased for the Centre and allied offices.

LAWG had always had a measure of financial and institutional support from denominational institutions in Canada and had engaged with religious orders and the international network of the Jesuit orders in its work. The rightward shift of the Roman Catholic church had a significant impact in Latin America and Canada. The gradual decline of "mainline" Protestant denominations in Canada involved a consolidation of the ecumenical social justice coalitions into one body – KAIROS – and a diminution in the availability of financial support to other organizations like LAWG. Acton and Anderson comment, "LAWG had difficulty finding its way forward in the post cold-war 1990s" (2017, 10). In March 1997 the decision was made at the annual meeting to formally wind up LAWG and find a suitable home for the resource collection.

The impact and legacy of thirty years of research, education, and action remained. As Brian Stevenson suggests in his extensive review of civil activism regarding the Americas, "without strong domestic pressures, there was little reason for Canadian governments to become as involved as they did in the politics of Latin America. Consequently, the Latin America constituency can be seen as the locus of Canada's foreign policy towards Latin America for over two decades" (2000, 222). As we have illustrated, the LAWG was at the heart of that constituency for close to three decades, organizing, informing, and influencing.

As for LAWG itself, one contributor commented that the decision was to "end LAWG as an organization, but not an idea." The LAWG history project (see below) slowly emerged.

NACLA and LAWG: A Reflection

As noted in the introduction, both LAWG in Canada and NACLA in the United States emerged in good part in response to the exercise of imperial control in the Western Hemisphere in the mid-1960s.

The critical factors that affected each organization differed. In the case of NACLA, questions of political discipline and collective process led to internal schisms which were complicated by the long-term maintenance of two locations, New York and the Bay Area. The East coast office continued to produce the highly respected *NACLA Report on the Americas*. In the West, led by NACLA co-founder Fred Goff, a new formation the "Data Centre: Research for Justice" was established in 1977, closing its doors in 2016. Building on previous history in corporate research and including a broader range of environmental and social justice themes, the centre developed a considerable history of projects informed by critical information.

The focus on data, on critical information about a host of Latin American societies and politics as well as on Canadian corporate and other interests continued to motivate LAWG veterans in establishing first CLARC and later, in collaboration with TCLSAC, a joint information library and research centre in Toronto. Separately incorporated from LAWG, Counterpoint Resource Centre sought to provide information and educational resources for a variety of clients. By 2001, the Centre was no more, and locations had to be found for its immense library of books, publications, and files.

Archive

When the collection of thirty years of the LAWG enterprise was inaugurated at CERLAC, tribute was paid to two people – York Professor Liisa North and former LAWG Librarian Caese Levo, who had negotiated the new home and curated the materials. LAWG was virtually unique among solidarity organizations in that it employed a librarian with professional qualifications for a number of years. Levo reflected on the experience as follows: "LAWG felt it was important that Canadians learn about what was really happening in the region. They focused their research and publishing on Canada's role – on Canadian foreign aid, foreign investment and the activities of Canadian corporations like INCO and Falconbridge" (Foster 2017).

Over the years LAWG members worked in the region and made connections with other organizations. They set up exchanges of publications and brought materials back from their travels. They subscribed to key journals like the *Latin America Weekly Report*, the *Central America Report from Guatemala* and the ISLA clipping service from Berkeley. The periodicals from the LAWG Library have been integrated into the CERLAC country and subject sections. The church, trade union, and human rights sections have especially grown as a result.

Two new sections have been created from the LAWG materials – one for Canadian solidarity with the different countries and the other for materials produced by Latin Americans in exile. Following military coups in the region, Latinos often found themselves in many different countries, and they created their own publications.

The LAWG Library reflects the focus of LAWG's research and solidarity work over the years so that you will find more material, for example, on the Dominican Republic, Chile, and Central America. LAWG worked closely with the Canadian churches, trade unions, and NGOs. The collection is unique because it has gathered together publications and ephemera from these sectors and preserved them.

As long-term LAWG Librarian Caese Levo remarked at the formal opening of the collection, "The richness of these materials is found in the country and subject files which are well organized here. When you look in this room you will see rows of boxes – but when you open those boxes you will hear the voices of those who worked for social justice in Latin America and also the stories of the Canadians who supported their struggles" (Levo 2017, 1; see also http://cerlac .info.yorku.ca/resources/collections/latin-american-working-group -lawg-library/).

New Life: Continuing the Narrative

Through reunions of former staff and volunteers in 2010, 2014, and 2015, a project in recuperation of history developed. The context was that of the Harper government and a sense of hostility to non-governmental organizations, generational change, and a potential loss of history and experience. A small group of former members initiated what was called *"Sí Hay Camino" The LAWG History Project,* which was completed in 2018. It was an enterprise of six volunteers supported initially by a part-time staff person. This work included the following:

- Assembling documents, publications, recordings, and photos to fill in gaps in the collection at York;
- Undertaking oral history interviews with a number of former staff and volunteers, which has become more urgent with the recent death of several;
- Initiating and supporting several further research investigations, on specific aspects of LAWG's work and on issues like the role of the security services and RCMP in surveillance of solidarity activities;
- Engaging our colleagues in identifying what we conclude were the top takeaways of the experience, items like action-oriented research; sustained collective commitment; trusting relationships; conjunctural analysis, etc.; and
- Establishing a website in which these new materials and findings can be found: https://lawghistoryproject.wixsite.com/lawghay caminohistory/the-project.

To return to the quotation with which I began – the creation of *narratives* and the evaluation of significance, the making of *history* – with the conclusion of the history project, the enterprise really moves to *other* hands. We look forward to the inquiries of other students, researchers,

activists, and academics and those who pick up the torches of solidarity, human rights, and social organization.

NOTES

1 This chapter makes extensive use of the work of Janice Acton, Betsy
   Anderson, Louise Casselman, Suzanne Dudziak, and Caese Levo. Useful
   background was found in the work of Dr. Bruce Douville of Algoma
   University and commentary by Prof. Luis van Isschot of the Department of
   History, University of Toronto. Thanks are due to Carleton student assistant
   Julia Van Drie.
2 While an employee and representative of the United Church of Canada,
   the author acted as chair of the ICCHRLA and for several months of the
   Toronto Welcome Committee for Refugees from Chile, as well as part of the
   governance bodies for the TCCR and GATT-fly.
3 The comment is credited to Texan Methodist Rev. Brady Tyson (Rosen 16).
4 For a thorough examination of this period and civic response, see Simalchik
   1993, see also Gunn 2018.
5 The development of a thorough economic analysis and critique of the post-
   coup Pinochet government in Chile as well as the role of development aid
   in poverty relief (or failure thereof) was contributed by LAWG researchers
   Robert Carty and Virginia Smith and volunteer John Dillon. The actions of
   Canadian banks and mining corporations were also critiqued.
6 LAWG co-founder John W. Foster became co-chair, with Professor Marjorie
   Cohen, of the Toronto Committee Against Free Trade, an alliance of labour,
   cultural, environmental, and feminist formations.
7 These initiatives are summarized in Foster 2005.

REFERENCES

Acton, Janice, and Betsy Anderson. 2017. *The LAWG Solidarity Story*. Accessed
   21 October 2021. https://lawghistoryproject.wixsite.com/lawghaycamino
   history/the-project.
Baranyi, Stephen, and John W. Foster. 2012. "Canada and Central America:
   Citizen Action and International Policy." In *Canada Looks South: In Search of
   an Americas Policy*, edited by Peter Mckenna, 240–64. Toronto: University of
   Toronto Press.
Brewin, Andrew, Louis Duclos, and David MacDonald. 1976. *One Gigantic
   Prison: The Report of the Fact-Finding Mission to Chile, Argentina and Uruguay*.
   Toronto: The Inter-Church Committee on Chile.

Carty, Robert, Virginia Smith, and LAWG. 1981. *Perpetuating Poverty: The Political Economy of Canadian Foreign Aid*. Toronto: Between the Lines.

CECOPE (Centro de Coordinacion de Proyectos Ecumenicos). 1989. *Proyecto "Fronteras Comunes Mexico-Canada."* Mexico: D.F.

CERLAC LAWG Library collection. Accessed 21 October 2021. http://cerlac.info.yorku.ca/resources/collections/latin-american-working-group-lawg-library/.

Common Frontiers. 1989. Project description, 22 September.

Deverell, John, and the Latin American Working Group. 1975. *Falconbridge: Portrait of a Canadian Mining Multinational*. Toronto: James Lorimer & Company.

Douville, Bruce Michael. 2008. "A Puppy-Dog Tale: The United Church of Canada and the Youth Counter-Culture, 1965–1973." Canadian Society of Church History. *2008 Historical Papers*.

– 2011. *The Uncomfortable Pew: Christianity, the New Left and the Hip Counter-Culture in Toronto. 1965–75*. PhD diss., York University.

Foster, John W. 2005. "The Trinational Alliance against NAFTA: Sinews of Solidarity." In *Coalitions Across Borders: Transnational Protest and the Neoliberal Order*, edited by Joe Bandy and Jackie Smith. Lanham and Oxford: Rowman & Littlefield.

– 2015. *Where LAWG Came from: From before Conception through Infancy: the Mid-to-Late Sixties*. Accessed 21 October 2021. www.lawghistoryproject.wixsite.com/lawghaycaminohistory.

– 2017. *Life Beyond Death: The Story of the Latin American Working Group*. Notes for a presentation to Carleton University Workshop: Canada's Past and Future in the Americas. 27–28 March 2017.

Graham, John. 2015. *Whose Man in Havana? Adventures from the Far Side of Diplomacy*. Calgary: University of Calgary Press.

Gunn, Joe. 2018. *Journeys to Justice: Reflections on Canadian Christian Activism*. Ottawa: Novalis.

Hamilton, Ian. 1970. *The Children's Crusade. The Story of the Company of Young Canadians*. Toronto: Peter Martin Associates Limited.

Harper, Charles R. 2006, *O Acompanhamento: Ecumenical Action for Human Rights in Latin America 1970–1990*. Geneva: World Council of Churches.

Imperialism & Change in Latin America. 1971. Toronto: Original program.

Kostash, Myrna. 1980. *Long Way from Home: The Story of the Sixties Generation in Canada*. Toronto: James Lorimer & Company.

LAWG. 1983. *Central American Women Speak for Themselves*. Toronto: LAWG.

LAWG History Project. 2018. "LAWG Hay Camino History Project." Accessed 21 October 2021. https://lawghistoryproject.wixsite.com/lawghaycaminohistory.

Lernoux, Penny. 1980. *Cry of the People: United States Involvement in the Rise of Fascism, Torture, and Murder and the Persecution of the Catholic Church in Latin America*. New York: Doubleday.

Levo, Caese. 2017. Remarks, LAWG/CERLAC Library Opening, Toronto, 10 February.

Lind, Christopher, and Joe Mihevc, eds. 1994. *Coalitions for Justice: The Story of Canada's Interchurch Coalitions*. Ottawa: Novalis.

Mass, Bonnie. 1976. *Population Target: The Political Economy of Population Control in Latin America*. Toronto: LAWG.

Matthews, Robert O., and Cranford Pratt, eds. *1988 Human Rights in Canadian Foreign Policy*. Kingston and Montreal: McGill-Queens University Press.

McFarlane, Peter. 1989. *Northern Shadows: Canadians and Central America*. Toronto: Between the Lines.

Mills, C. Wright. 1956. *The Power Elite*. New York: Oxford University Press.

– 1960. *Listen Yankee: The Revolution in Cuba*. New York: Ballantine Books.

Moffatt, Gary. n.d. *History of the Canadian Peace Movement until 1969*. Toronto: n.p.

North, Liisa. 1985. *Negotiations for Peace in Central America: A Conference Report*. Ottawa: Canadian Institute for International Peace and Security.

North, Liisa, and CAPA, eds. 1990. *Between War and Peace in Central America: Choices for Canada*. Toronto: CAPA.

Ogelsby, Carl, and Richard Shaull. 1967. *Containment and Change*. New York and Toronto: Macmillan.

Park, L.C., and F.W. Park. *Anatomy of Big Business*. 1962. Toronto: Progress Books.

Project Description. 29 September 1989. Common Frontiers. Stage II.

Project Brazil, and Last Post Staff. 1973. "The Brascan File: Its Friends in Government, Its Record in Brazil." Toronto: The Last Post.

Randall, Margaret. 1974. *Cuban Women Now: Interviews with Cuban Women*. Toronto: The Women's Press.

Rosen, Fred. 2002. "NACLA: A 35 Year Retrospective." *Report on the Americas*. Vol. XXXVI, No. 3 New York: NACLA.

Rowntree, John, and Margaret Rowntree. 1968. *The Political Economy of Youth (Youth as Class)*. Ann Arbor, MI: The Radical Education Project.

Simalchik, Joan. 1993. *Part of the Awakening: Canadian Churches and Chilean Refugees*. Master's thesis. University of Toronto.

Stevenson, Brian J.R. 2000. *Canada, Latin America, and the New Internationalism: A Foreign Policy Analysis, 1968–1990*. Montreal and Kingston: McGill-Queens University Press.

Trouillot, Michel-Rolph. 1995. *Silencing the Past: Power and the Production of History*. Boston: Beacon.

4 Canadian Security and Defence Policies towards Latin America: Liberal Engagement or Harsh Realism?

FEDERMÁN RODRÍGUEZ

Canadians have a split personality. We seem constantly attracted to opposite poles in our thinking about our role in the world. One pole ties us to hard reality, *realpolitik* if you will, and makes us want our governments to protect the national interest when it deals with other states. Canadians, when they think this way, talk in terms of our sovereignty, security, territory, trade, economic growth and prosperity. In contrast to the pole of realism, there is another pole that attracts Canadians to an idealistic vocation. Its advocates tend to have a visionary, at times almost romantic, approach to our position in the world
Alan Gotlieb (2014). Canadian Ambassador
to the United States (1981–9)

Canadian foreign and security policies (CFSP) towards Latin America seem to embody Alan Gotlieb's thesis about Canadians' split personality and their subsequent involvement in the world. On the one hand, Canadian engagement with security and defence issues in the region has been certainly the result of a liberal, broad, and, even romantic, approach to security. Not only has this approach led to Canadian involvement in peaceful resolution of disputes and confidence-building initiatives, such as those advanced during the Central American civil wars in the 1980s but it has also resulted in human security– and multidimensional security–oriented agendas during the 1990s and 2000s, respectively (Klepak 2012, 42–3; Mace et. al. 2012, 607–19). More recently, under similar motives, Haiti became the largest beneficiary of development assistance and the clearest example of the security-development nexus within CFSP in the Americas (GC 2009, 15; Baranyi 2014). Since Canada joined the Organization of American States (OAS) in January 1990, such a liberal approach has also entailed support for updating the hemispheric security institutional apparatus to make it more suitable to the post–Cold War threats and challenges (Mace et al. 2012, 611). From this perspective, the Harper government's Americas Strategy seems to be the natural conclusion of a long journey towards the refinement of CFSP, in which security is inextricably linked to democracy and prosperity as the three main goals of Canada in the region. Furthermore, the Trudeau government's renewed internationalism would offer an enticing prospect for continuing advancing Canadian broad engagement in Latin America.

On the other hand, a closer examination shows Canadian involvement in security and defence issues in Latin America has also at times been derived from a realist, narrow, and even relentlessly pragmatic approach to security. Not only did this approach result in a calculated and selective engagement during the Cold War to prevent Canada from being immersed in the American fight against communism, as discussed in the chapter in this volume by McKercher, but also the security agenda in the early post–Cold War period seemed to respond to, or even to be subordinated to, the search for commercial markets in the region (Klepak 2012, 33; Mace et al. 2012, 609). After all, the country's membership in the North American Free Trade Agreement (NAFTA), which brought Mexico into the US–Canada free trade agreement, constituted a springboard for a more robust agenda, including security and defence issues (see Morden 2012; Benitez and Hristoulas 2012). Furthermore, despite being a full-fledged member of the OAS and having promoted the Inter-American Convention against Terrorism after the 9/11 attacks, Canada eschewed embracing those hemispheric institutions concerning collective security commitments, such as the Rio Pact (Klepak 2012, 36). After all, the only genuine Canadian commitment to collective security has been the North Atlantic Treaty Organization (NATO) (Jockel and Sokolsky 2009, 317). Canada has also not ratified the Inter-American Commission on Human Rights, which has prevented it from operationalizing broad conceptions of security. Building on this account, the Harper government's Americas Strategy can be seen more as a strategy aimed at seizing considerable economic-security opportunities than as a robust commitment with the region. Under such circumstances, despite having promised a broad agenda regarding security and defence, the Trudeau government seems to have followed the same economic-security path followed by their predecessors in the Americas.

Both of these approaches to security are often hardly distinguishable in addressing CFSP. After all, liberal values and strategic interests are two sides of the same coin when the foreign policies of democratic-liberal countries come under scrutiny. However, this chapter argues that such policies must be broken down into their liberal and realist components to understand not only their actual reach but also their potential contradictions, failures, and possibilities for improvement. By liberal security policies, I understand those that make individuals' security around the world the centre of its objectives, as well as those aimed at engaging with global issues as a result of a cosmopolitan duty. By realist security policies, I mean those that make the defence of national interests the centre of its objectives, which implies defence measures within the national and international realms. In considering economic security as being part of realist security policies, I adopt Barry Buzan's definition, namely, the "access to the resources, finance and markets necessary to sustain acceptable levels of welfare and state power," which implies, at the same time, guaranteeing the needed security conditions to have such access (Buzan 1991, 433).

Drawing on critical works aimed at addressing the interaction between values and interests within CFSP (Melakopides 1998; Nossal 2007; Gotlieb 2004), I explain, in the first section, how Canadian engagement with Latin America during the Cold War and the early years of the post–Cold War managed to harmonize liberal values and strategic interests within what Costas Melakopides and Kim Richard Nossal called *pragmatic idealism* and *liberal realism* to explain the broad Canadian foreign-policy objectives. However, values and interests become divorced from each other as economic security interests related to the commercial ties with the United States and energy and mining sectors in the Americas have emerged during the post–Cold War. That is to say, economic security interests have eclipsed, or even undermined, liberal policies towards the region. This makes Gotlieb's thesis of a Canadian split personality a compelling explanation of CFSP towards Latin America and the Caribbean during this period. In the second section, I review the central controversies related to the security dimensions of the Harper government's Americas Strategy, as well as how such a strategy might be the best recent example of the divorce between liberal values and strategic interests. As concluding remarks, I examine whether the Trudeau government's foreign and security policies can pave the way for a broad engagement willing and able to harmonize values and interests or if it constitutes another split engagement with the region. Certainly, from the perspective of Latin American countries, a broad conception of security and defence would be critical to breaking up the unequal and violent structures of power which characterize some segments of societies in this region (see Hilgers and Macdonald 2017, 10).

On Values and Interests

In attempting to explain the history of CFSP towards Latin America and the Caribbean, three phases of international engagement can be highlighted, namely, pragmatic idealism during most of the Cold War, liberal realism during the transition to the post–Cold War, and the split involvement during the post–Cold War as economic-security interests started to undermine liberal engagement. Explaining these phases enables us to understand respectively how Canadian values and interests have been harmonized, how tensions remained between them or how they have recently been contradictory.

Pragmatic Idealism during the Cold War

During most of the Cold War, Canada's involvement in the region was informed by pragmatic idealism. As a world view, this entailed not only embracing the values of justice, the meeting of human needs and rights worldwide, the achievement of duties beyond borders, moderation, multilateral cooperation, and generosity, but also adjusting such values to national interests according to political circumstances (Melakopides 1998, 4–6). Two initiatives were particularly emblematic of pragmatic idealism, namely, Canadian policies towards post-1973 Chile and the Central American civil wars during the 1980s. Firstly, even though Canada recognized the Chilean military junta headed by General Augusto Pinochet two weeks after the overthrow of the democratically elected socialist government of Salvador Allende on 11 September 1973, the government of Pierre Trudeau accepted about 7,000 Chilean refugees when the new military regime oppressed the political opposition and perpetrated systematic human rights violations. This phase would forever change Canadian refugee policies at the service now of pragmatic idealism (See Stevenson 2000, 122–31).

Secondly, despite its growing interest in strengthening ties with the United States during the 1980s, Canada undertook and supported diplomatically and economically different peace initiatives towards Central America, such as the Contadora Plan. Such initiatives took issue with the Reagan government's "solution" for the Central American civil wars based on fighting against communism relentlessly. The Canadian government sought to minimize the exacerbation of such wars as a result of the controversial support from the United States for the *contras* in Nicaragua, and the military-led governments of El Salvador and Guatemala. Unlike such typical Cold War strategies, Canada recognized at the diplomatic level that the central problems of the region were rooted more in social and economic conditions than in the expansion of communism. Such a pragmatic idealism was operationalized through total direct development assistance to Central America of about $170 million in 1987. Furthermore, continuing its 1970s refugee policy, Canada supported the UN High Commission for Refugees and the International Committee of the Red Cross and admitted about 21,000 refugees from Central America (Melakopides 1998, 150–1; see also Stevenson 2000, 147–52).

However, the adoption of refugee policies and peace initiatives did not imply a genuine and active involvement regarding security and defence in the region. The fear of being immersed in the Cold War and, accordingly, entangled with American strategies against communism, as well as of having to take sides within US–Latin America disputes, constrained significantly the Canadian government's leeway (Rochlin 1994; Klepak 2012; Mace et al. 2012). After all, after the Cuban Revolution and the subsequent replacement of the government of Fulgencio Batista by the first revolutionary socialist regime in the region, the United States backed military coups and authoritarian governments and adopted some counter-insurgency strategies in the region to avoid the expansion of communism (Klepak 2012, 33; Mace et al. 2012, 609) (not to mention American governments' attempts to promote its economic agenda). Likewise, Canada was hesitant to establish closer ties with some Latin American states because of the predominance of military regimes. As a result, Canada's limited agenda in the region focused more on economic and diplomatic objectives than on security and defence issues, as was displayed during the first ministerial mission to the region in 1968 (Rochlin 1994, 77 and Chapter 5).

Liberal Realism during the Transition to the Post–Cold War

During the transition from the Cold War to the early post–Cold War, it can be argued that Canada embraced *liberal realism* in conducting foreign and security policies towards the region. Seen as a more refined version of Cold War pragmatic idealism, it implied the need to "understand and accept the importance of power in determining outcomes of conflicts between peoples, communities, and states [as well as] the necessary diversity and heterogeneity of political communities in the contemporary international system" (Nossal 2007, 269). Above all, it implied eschewing the *sanctimonious idealism* that would become dominant during the Chrétien and Martin governments at the global level. It is worth noting that this latter form of idealism was based on "self-righteous pretension to a superior morality," that is, the promotion of universal values rather than heterogeneous conceptions of rightness and justice (265–7).

The end of the Cold War, as well as the expansion of democratization and economic liberalization processes, and the decline of the US hegemony in the region seemed to pave the way for a deeper Canadian involvement through liberal realism (see Macdonald 2016, 5). Above all, it was operationalized through the stationing of defence attachés in embassies,[1] the involvement in peaceful resolution of disputes and confidence-building initiatives – echoing its experience during the Central American civil wars in the 1980s – the promotion of democracy, the coordination of peace-keeping operations along with its Latino partners, and the expansion of the Military Training and Assistance Program (Klepak 2012, 42; McKenna 2018, 29). To be sure, such measures were in line with liberal realism as all of them recognized the transformation of international order and avoided imposing a Canadian values–oriented view on the region.

The promotion of broad conceptions of security, usually associated with sanctimonious idealism (Gotlieb 2004; Hampson and Oliver 1998, 379–406), was even adapted to the regional needs and, accordingly, to liberal realism's insights. For instance, the concept of human security introduced by the 1994 UN Development Report was not only one of the primary pillars of the Chrétien government's foreign and security policies, but also one of the guiding principles of its so-called defence diplomacy towards the region (Klepak 2012, 43; DFAIT 1995; Bosold and Von Bredow 2002, 837, see also Tomlin 2008, et al. 222–9). The campaign to ban landmines, which resulted in the Ottawa Treaty, one of the primary Canadian human security initiatives, was possible in large part thanks to the ability of Minister of Foreign Affairs Lloyd Axworthy to obtain the 100 per cent support of Latin American countries (Klepak 2012, 43). On the other hand, despite its commitment to traditional conceptions of security within the War on Terror after the 9/11 attacks, Canada seized the opportunity offered by the 2003 Declaration on Security in the Americas and mainly its concept of multidimensional security to continue adopting a broad engagement in the region. In fact, during the 2000s, Canada not only engaged with the Inter-American Convention against Terrorism but also continued with its active military cooperation, the fight against drug trafficking – usually linked to human and arms trafficking and money laundering – and worked closely with the Pan American Health Organization and the Pan-Caribbean Public Health Agency to deal with health and natural catastrophes as threats to hemispheric security (Mace et. al. 2012, 614–18).

The Emergence of Economic Security Ambitions

The relative harmonization between interests and values, which was achieved by the Cold War pragmatic idealism and the early post–Cold War liberal realism, has been progressively undermined by the emergence of an economic security sub-agenda towards the region. This approach was initiated by the Chrétien Liberals and deepened by the Harper Conservatives.

Even though Canada held that the disappearance of a bipolar order encouraged broad security agendas aimed at acknowledging and facing traditional and non-traditional threats, its agenda towards Latin America has involved a growing economic security lens. This agenda was justified due not only to domestic fiscal constraints but above all to the progressive transformation of power in the international realm, which was defined increasingly in economic terms (DFAIT 1995, Introduction and chapter IV; DND 1994, chapter 1). Canada's engagement first with NAFTA and then with the Free Trade Area of the Americas and finally with the signing of bilateral free trade agreements with Chile (1997), Costa Rica (2002), Peru (2009), Colombia (2012), Panama (2013), and Honduras (2013) would be good examples of such an approach (see Macdonald 2016, 5 and 7).

More notably, the Canadian economic-security agenda was arguably advanced through a rationale expressed as follows: The need for prosperity in a highly globalized world encourages Canada to contribute to the proper functioning of the global economy, as well as to the preservation and maintenance of international peace and security. That is to say, an export-oriented economy like Canada's that depends on global stability must contribute to such stability. Previous Canadian governments had taken into account both prosperity and national security as critical foreign-policy objectives. However, unlike its predecessors, the Chrétien government seemed to view prosperity as a matter of security (Rodríguez 2017, 51–3). The Chrétien government clearly expressed this view in *Canada and the World: Canadian Foreign Policy Review 1995*:

> Our own security, including our economic security, is increasingly dependent on the security of others. More than ever, the forces of globalization, technological development and the scale of human activity reinforce our fundamental interdependence with the rest of the world…. Canadians recognize the vital link between their own security and prosperity and the security of others. Just as Canadians appreciate that prosperity demands the best possible mix of domestic and international economic policies, so

too they realize that protecting and enhancing their security and prosperity requires a security policy that promotes peace in every part of the world with which Canada has close economic and political links (DFAIT 1995, summary and chapter IV).

On the surface, pragmatic idealism could have been the main driving force of such an approach to security as Canadian interests were intertwined with global concerns on security and prosperity in a pragmatic way. Certainly, enhancing security and prosperity implied a cosmopolitan engagement solely in "every part of the world with which Canada has close economic and political links." This agenda could have even embraced liberal realist insights since a cosmopolitan engagement did not entail imposing Canadian liberal values. However, during the post–Cold War era, Canada's economic security agenda not only started to be consolidated as a new component of the broad approach to global security but has even contradicted liberal yearnings. In particular, the insatiable search for markets in the energy and mining sectors led Canada's ambivalent engagement with human rights in the region, as discussed in the chapter by Jen Moore in this volume (Shamsie and Grinspun 2010, 188; Macdonald 2016, 7).

Together with the Canadian tendency to conduct its liberal and economic-security engagements in Latin America in a contradictory manner, another stance has been to split economic and security issues in ways that can undermine the government's broad engagement with the region. That is to say, Canada has distinguished its economic-security agenda from other security and defence issues in an attempt to protect its strategic interests and, above all, to deal with the US factor in the region. As a result, its engagement has at times ended up being narrower than it has officially held. Canada–Mexico relations during the post–Cold War era are the most salient example of such a stance. On the surface, NAFTA seemed to imply a deep integration between Canada, the United States, and Mexico. The free trade agreement brought about the integration of commodities, investment, production, and (to a limited extent) labour markets. The North American region also adopted robust security agenda, including economic security, military security, food security, energy security, environmental security, biosecurity, and, after the 9/11 attacks, anti-terrorism. The signing of the Security and Prosperity Partnership in 2005, the collaboration between health ministers and scientists to deal with the H1N1 epidemic and the talks about security perimeter or NAFTA-Plus seemed to be, among several initiatives, the primary markers of such an allegedly sturdy agenda.

However, as some observers have argued, Canada has considered from the very beginning of the integration process that the inclusion of Mexico in the North America agenda is not in its best interest (Benítez and Hristoulas 2011, 223). Such fear was particularly true after the 9/11 attacks and the subsequent anti-terrorist agenda advanced by the United States. In particular, the Martin government considered that with a security perimeter for North America to prevent new terrorist attacks, Canada would have not only had a reduced room for manoeuvre in critical areas, such as asylum and refugee policies, but, above all, would have had to accept the same conditions imposed on Mexico to manage its border security. After all, the threats presented at the US–Canada border are profoundly different from those at the US–Mexico border. Also, despite continuing defending trilateral agreements at the economic level, the Martin government decided to strengthen bilateral security agreements with the United States (Sokolsky and Lagassé 2005/2006, 20–1). This strategy continued the Chrétien government's decision to sign the bilateral Canada–US Smart Border Declaration in 2001 aimed not only at cooperating regarding border security but also at preventing the disruption of bilateral trade (see Clarkson and Banda 2003, 5; Sands 2008, 110; Burt 2009, 168).

To summarize, in attempting to address the relationship between values and interests within CFSP towards Latin America up to the 2000s, it seems that pragmatic idealism, the country's dominant approach during the Cold War, and liberal realism, advanced in the transition to the post–Cold War were replaced by a split and contradictory stance between a broad liberal commitment and an economic security engagement during the post–Cold War period. The reason was that Canadian growing strategic interests began to clash with the promotion of liberal values in the region. As explained in the next section, this became most obvious during the Harper government.

The Harper Government's CFSP towards Latin America

The Harper Conservatives officially made Latin America and the Caribbean a top international priority through the Americas Strategy announced in the spring of 2007. This strategy embraced, at first sight, a broad liberal engagement with the region. Advancing a common security agenda was one of its strategic objectives, which was inextricably linked to the other two, namely, increasing economic prosperity and reinforcing democratic governance. Officially, the underlying motive of the strategy was that Canada is a country of the Americas, for geographical and historical reasons, as well as because of its multidimensional ties "in terms of trade, immigration and cultural and social exchanges" across the region (GC 2009, 4).

At home and abroad, however, there were huge controversies about the real motives for the strategy as well as about the government capacity to progress from words to deeds, especially regarding the availability of economic and bureaucratic resources, and the government's ability to intertwine security yearnings with economic interests. After explaining the primary hypotheses about the "real" motivations, and inconsistency of the Americas strategy, I examine how the Canadian split personality has brought about a complex, even contradictory, security and defence agenda.

Americas Strategy's Controversies

The first controversy is related to the strategy's real motivations. In this scenario, three hypotheses, regarding domestic politics, the identity of Canadian foreign policy and the changing international order, are particularly palpable in the security and defence realms. The first hypothesis points out that Canada redirected its attention towards the region to meet the Harper government's political constituencies and domestic constraints. On the one hand, as discussed in Jen Moore's chapter, Canadian mining interests have increased significantly over the last three decades as Canadian companies have managed to dominate almost the half of the mineral exploration market in the region. Since many of these companies are from Western Canada, which was the primary constituency supporting the Harper government, it can be argued that Canadian economic-security engagement in the region has operated at the service of corporate constituencies (Shamsie and Grinspun 2010, 186; McKenna 2012, xv). On the other hand, Canadian security and development aid to Haiti seemed to be the result of electoral calculations, even after the 2010 earthquake, due to the significant political clout of Quebec's Haitian diaspora and particularly its role in several Montreal ridings (Shamsie and Grinspun 2010, 185; Macdonald, 2016a,11).

The second hypothesis suggests that the Harper Conservatives were interested in rebranding Canadian contributions to global security and prosperity by clearly differentiating themselves from the Martin Liberals. After all, the Martin government had made Africa one of their primary priorities, along with Afghanistan, and had brushed aside Latin America in their official policy statements (McKenna 2018, 23; Macdonald 2016, 6 and 10). Above all, during this government, Canada had developed an integrated 3D approach aimed at combining diplomacy, defence, and development strategies mainly towards fragile and failed states in Africa (GC 2005a, 20). Within the Martin government's international policy strategy, Haiti was the only Latin American country which seemed to receive the same attention in terms of the 3D approach and particularly in terms of security (GC 2005d, 5). What is more, if we considered development aid as an extension of Canada's broad security involvement, Latin America would not have been yet a priority for the Liberals. Of twenty-five countries prioritized in the development assistance policy, only four were from Latin America, namely, Bolivia, Guyana, Honduras, and Nicaragua (Shamsie and Grinspun 2010, 185).

According to this hypothesis, during the Harper government, international assistance priorities seemed to change. Without losing sight of Afghanistan, which was the target of a massive and unprecedented Canadian intervention (Moens 2007), Canada seemed to draw attention towards Latin America to rebrand its contribution to international security through international assistance (see McKenna 2012, xiv). Beginning in 2009, through its Aid Effectiveness Agenda, the Harper Government made it clear that bilateral resources within its top twenty recipient countries would be destined now to critical Latin American recipients, namely, Bolivia, Colombia, Haiti, Honduras, Peru, and the Caribbean Regional Program (Shamsie and Grinspun 2010, 185).

However, this hypothesis suffers from an important flaw. In considering international assistance to Latin America and Africa in aggregate and historical terms, it can be argued that Latin America has never been a priority over Africa, regardless of whether Conservatives or Liberals come to power. Figure 4.1 showcases this by considering the evolution of spending allocated by the Canadian International Development Agency (CIDA), the Department of Foreign Affairs and International Trade (DFAIT), the Department of National Defence (DND), the International Centre for Human Rights and Democratic Development (ICHRDD), and the Royal Canadian Mounted Police (RCMP) to international assistance towards Africa and Latin America. The spending of these agencies, whose efforts can be considered together to understand CFDP towards Latin America, also evolved according to the patterns of the total spending on international assistance. Above all, these figures indicate that Africa and Latin America have kept their relative weight within CFDP over the last three decades. The international assistance to both of these regions increased progressively and simultaneously between 2001 and 2012, from $257 to $893 million towards Latin America, and from $515 million to $2,085 million towards Africa. It can be argued even that the Canadian government prioritized Africa increasingly as a percentage of the total spending during these years.

Figure 4.1. Canadian International Assistance (CAD$M)

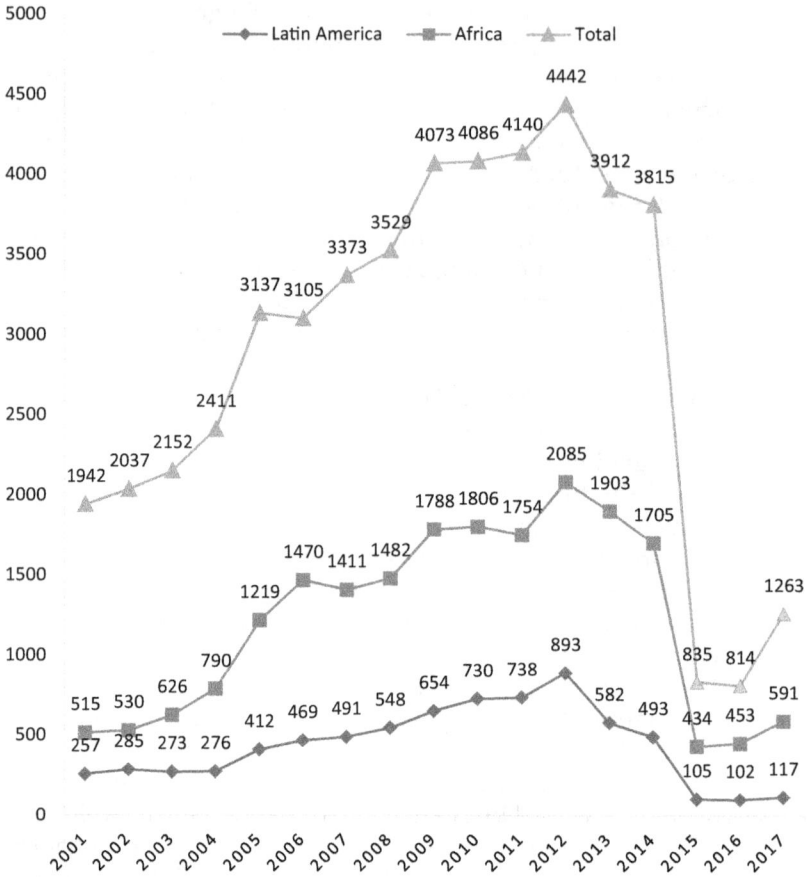

Data accounts for the sum of spending allocated by CIDA, DFAIT, DND, ICHRDD, and RCMP.
Source: CIDP 2020.

While international assistance to Africa increased from 26.5 per cent to 46.9 per cent, international assistance to Latin America only increased from 13.2 per cent to 20.1 per cent. Not surprisingly, the steady and slight decline from 2012 to 2014, as well as the sharp decrease in 2015 in the total spending impacted the Canadian government's priorities towards Latin America and Africa in percentage terms as well. [2]

The third hypothesis maintains that Canada's re-engagement regarding security and defence towards Latin America can be seen as the result of international constraints. Some would argue that in a post-hegemonic international order in which the scant room for manoeuvre for the United States was palpable in the region, the Harper government's aspiration for strategic repositioning at the global level could have affected its interest in adopting a growing strategic role in the region. To refine such hypothesis, however, it is indispensable to acknowledge that the Harper government managed to consolidate its ties only with right-wing and conservative governments in a divided region which had seen the emergence of counter-hegemonic regional blocs, such as the Union of South American Nations (UNASUR) and the Bolivarian Alliance for the Americas (ALBA). Such blocs excluded not only the United States but also Canada (see Macdonald 2016, 4). In light of this hypothesis, it can be held, in the final analysis, that Canada's re-engagement in Latin American security and defence issues during the Harper government "was less a function of domestic politics ... and more the result of external stimuli," whether the decline of US hegemony or regional transformation (McKenna 2018, 33).

The Americas Strategy's second controversy was related to the Harper Conservatives' ability to progress from words to deeds beyond meeting their economic security interests in the energy and mining sectors. Despite promising to play a more prominent role in the Americas through a consistent and longstanding engagement, many observed that the Americas Strategy ended up being a set of hollow words (Miller 2018; McKenna 2018; Macdonald 2016). Besides avoiding the adoption of specific actions and initiatives aimed at operationalizing the silos of democracy, prosperity, and security, Canada did not allocate sufficient financial and bureaucratic resources to underpin them (McKenna 2018, 23 and 25).

At first sight, Canada attempted to adopt a more coordinated whole-of-government approach to guarantee a new level of engagement through the creation of a post called the Minister of State for Foreign Affairs in 2008 and high-level visits by officials to the region. The former was instrumental in displaying a new strategic and comprehensive approach to the region, whereas the latter paved the way for the signing of several bilateral free trade agreements. However, the Secretary of State post was eliminated in 2013, probably as a result of a strategic shift towards Asia. For its part, the high-level visits were more episodic moments within the history of Canada–Latin America relationships than the beginning of a calculated, sustained, long-term, and continuously evaluated plan to the region. Without a coordinated strategy and long-term plan in mind, government departments adopted their own approach to the Americas, without investigating what Latin Americans really wanted when security and their correlated conditions, prosperity, and democracy, come under scrutiny (Macdonald 2016, 6–7, see also McKenna 2018, 25). In the final analysis, the Harper Conservatives, according to this perspective, can be seen as part of a long series of Canadian leaders, starting with the Diefenbaker Progressive Conservatives and Trudeau Senior Liberals, aimed at (re)discovering and consequently engaging with the Americas but without achieving significant concrete deeds (McKenna 2018, 25).

Although these controversies and hypothesis help clarify the nature of the Americas Strategy, I believe it is important to move beyond them to look at the impact of the two driving forces, liberal engagement and relentless realism, that have informed Canadian foreign policy in general, and CFSP towards Latin America and the Caribbean in particular. Specifically, I argue the split personality thesis can contribute extensively to the debate about the motivations of the Americas Strategy, as well as help explain its inherent inconsistency.

A Liberal and Broad Engagement

At first sight, Harper's vision in the Americas could look like a dream come true from the perspective of hemispheric organizations, several Latin American governments, political parties, and civil society organizations. After all, the Harper government not only made an official commitment to the three central concerns in the region, namely, security, democratic governance, and prosperity, but also considered them as "three interconnected and mutually-reinforcing objectives" (GC 2009, 6). Regarding security, Canada considered that regional stability depended on facing critical threats, such as "drugs, organized crime, health pandemics and natural disasters." Regarding prosperity, Canada promised to help the building of "dynamic and growing economies," as well as to "promote responsible investment and open markets that can create new opportunities." And as for democracy promotion, "freedom, human rights and the rule of law" were purportedly the guiding values informing Canadian engagement, with an emphasis on the strengthening of democratic institutions (GC 2009, 6). What is more, the Harper government's security and defence agenda in the region can be broken down into different components aimed at covering different types of threats, namely, traditional threats, such as drug trafficking and related crimes and violent crimes; post–11 September threats, such as terrorism and border insecurity; and new threats, such as health and natural catastrophes.

As for drug trafficking and related crimes – such as human and arms trafficking and money laundering – and other violent crimes, Canada not only encouraged the adoption of the Hemispheric Plan of Action against Transnational Organized Crime, but also joined the OAS working group charged with implementing it (Mace et al. 2012, 618). Such anti-crime efforts went hand in hand with its active role in the OAS Inter-American Drug Abuse Control Commission (CICAD) and the United Nations Office on Drugs and Crime (UNODC). Furthermore, the Harper government made efforts "to provide training and technical assistance aimed at improving international legal cooperation, and support mentoring programs and criminal justice reforms." (GC 2009, 13).

Concerning those programmes, Canada's involvement in Haiti, facilitating the United Nations Stabilization Mission in Haiti (MINUSTAH), can be considered one of the most paradigmatic examples of Canadian broad engagement in the region, mostly in terms of the fight against violent crime. Canada's role in that country is, however, the subject of considerable controversy. Among opponents, Kelvin Walby and Jeffery Monaghan (2011) argued that government agencies, such as RCMP, CIDA, and the Correctional Service of Canada (CSC), securitized the violence in Haiti and, accordingly, treated it as a direct existential threat to Canadian interests. Such an analysis could help understand how the realist component undermined the liberal component of the Harper government's involvement in Haiti. Although acting under the umbrella of the Stabilization and Reconstruction Taskforce (START), which aimed to adopt a broad approach to security and create synergy between agencies dealing with "the complexities of fragile states and international crises" (GC 2016), Canada focused, as mentioned by Walby and Monaghan (2011, 277–9), solely on critical policing measures, the construction and refurbishment of prison and jails, and border security.

From a different perspective, Stephen Baranyi (2014, 165) makes a case for a broad engagement thesis by arguing that Walby and Monaghan ignored the limited role of security and policing reform assistance vis-à-vis broad Canadian objectives. These refer to "humanitarian aid and cooperation for long-term economic, political and social development" (165). Baranyi argues that Canadian assistance, with all its deficiencies, may have been at the service less of securitization project and more of balancing policing and prison reform initiatives with a broad international assistance engagement. This was true even though these reforms advanced modestly, and the Haitian judicial system could not appropriately supplement police reform (168). What is more, it cannot be forgotten that police, judiciary, and prison systems' transformation into "professional and accountable institutions" was seen within the democratic governance component of the Americas Strategy; that is to say, they were regarded as "the building blocks of democratic governance"

(GC 2009, 9). Not surprisingly, during the period 2006–10 only $110 million of $840 million (13 per cent) of Canadian international assistance to Haiti was allocated for the policing and security reforms discussed by Walby and Monaghan. The remaining 85 per cent was focused on a broad project of reconstruction and humanitarian and development aid (Baranyi 2014, 169). That is to say, CFSP towards Haiti could be interpreted more from the perspective of a liberal, broad engagement aimed at combining security and development and humanitarian aid, rather than of narrow, relentless realism focused on containing violence inside Haiti (see Baranyi 2014, 166–70).

Regarding post–11 September threats, under the Harper government, Canada was one of the most active OAS members in advancing the fight against terrorism in the region, reflecting its close ties with the United States. Besides making the Inter-American Convention against Terrorism a point of reference of its CFPS towards the Americas, Canada, during the Harper government, became the most significant contributor to the Inter-American Committee against Terrorism (CICTE). In 2010, its contribution, which reached $1.5 million, supported several anti-terrorist projects regarding "cyber security, port security, bioterrorism, simulation training, and anti-money laundering" (Mace et al. 2012, 618). Such efforts can also be understood within the context of traditional security and defence cooperation with the region, which has driven Canada to participate actively in the conferences of American armies and defence ministers following the establishment of five defence attachés to manage bilateral defence relations with twenty-two countries in the region (GC 2009, 14).

Likewise, military and police training were an essential element within the Americas Strategy. Since 2008, Canada provided "professional development and peace support training ... to twenty countries in the region [which account for a] third of cooperation programme budget of $20 million went to the Americas" (Mace et. al 2012, 616). Those efforts were supplemented by stabilization and reconstruction tasks. Canada spent $1.5 million in the Latin American capacity-building programme and its successor the Latin American peacekeeping partnership since 2009. Haiti and Colombia were the primary beneficiaries of the stabilization and reconstruction task force, with $75 and $15 million, respectively, for "conflict prevention and peace-building efforts" from the period from 2006 to 2010 (GC 2009, 15; see also Mace et. al. 2012, 617). In particular, Canada directed significant efforts to support the OAS Mission to support the peace process in Colombia (GC 2009, 10; OAS 2016).

Finally, within the Americas Strategy, natural disasters and epidemics were seen as security issues. The 2010 earthquake in Haiti and successive hurricanes and floods affecting the region made the Harper government aware of how such disasters can not only hinder or even reverse some Latin American countries' development but also constitute a direct threat to public security. Haiti's earthquake and its aftermath were probably the clearest examples of this sad reality. Not only did it destroy several facilities and prisons, "allowing 4000 prisoners to escape and causing the death of hundreds of police officers," but it also stimulated the increase of "homicides, property crimes and sexual violence" (Baranyi 2014, 167). With this rationale in mind, natural disasters were seen as a challenge within the security component of the Americas Strategy, and Haiti became the largest recipient of foreign aid in the Americas, reaching a peak in 2011 with $354 million (GC 2009, 14; CIDP 2020). The Harper government also made preventive and health-related initiatives a vital goal within its inter-American security agenda. It aimed to support the creation of the Pan-Caribbean Public Health Agency (CARPHA) to integrate and coordinate existing health institutions, which was added to its efforts in combating dengue fever in the region (GC 2009, 14).

A Conservative and Economic Security–Driven Agenda

During the Harper government, Latin America became a priority concerning economic security as well. Strong commercial ties seemed to leverage the whole strategic interest in the region. Trade between Canada and the Americas increased from about $40 billion in 2006 to $56 billion in 2014, some major companies, like Scotiabank, Bombardier, SNC-Lavalin, and Brookfield Asset Management, saw the region as market niches for their future economic growth and, above all, large and small Canadian companies established a lucrative business in much of the region (McKenna 2015; Miller 2018, 1). During the time Harper Conservatives were in office, "Canadian mining companies' investment in Latin America represented more than 60 percent of total mining investment in the region, and more than half of Canadian mining investment worldwide" (Macdonald 2016, 8). Not surprisingly, the Harper government also decided to continue advancing, as mentioned, FTAs with several Latin American countries (See GC 2009, 11; Macdonald 2016, 7).

At first glance, the Americas Strategy seemed to combine such legitimate economic security aspirations with a liberal and broad engagement in the region. After all, Canadian contribution to security and democratic governance in the region could have offset its current and prospective lucrative benefits. Even in the security economic realm, this strategy seemed to embrace a win-win partnership with Latin America (GC 2009, 11–12). However, since such aspirations were built on conservative ideological leanings and were at times advanced regardless of their negative impact on human rights, they undermined, or even contradicted, the broad engagement mentioned above. On the one hand, the focus countries for development assistance, except for Haiti, were countries which had signed FTAs with Canada. These countries were not only middle-income developing countries but also shared neoliberal approaches to development and at times conservative leanings regarding security; that is to say, militarized approaches to the fighting against terrorism, drugs, and guerrillas, such as the case of Colombian President Álvaro Uribe's two governments (2002–6, 2006–10) (see Government of Colombia 2003; McKenna 2018, 24).

What is more, during the Harper government, many of them – mostly the Pacific Alliance members (Chile, Colombia, Mexico, and Peru) – represented the counterpoint of the Left turn in Latin America, as represented mostly by Nicaragua, Ecuador, Venezuela, and Bolivia and, to a lesser extent, by Brazil, Argentina, and Uruguay. By signing FTAs and subsequently increasing the international assistance programmes, including security and defence support, with the right-wing and conservative governments in the region, the Harper government took sides in the hemispheric game. This strategy seemed to vitiate its desire to contribute to security and defence from a broad perspective, which would arguably have entailed dealing with the ideological diversity in Latin America and the Caribbean.

On the other hand, the Harper government's economic-security agenda in the region led it to be ambivalent regarding human rights violations in the region. The case of Colombia is probably the most paradigmatic example in this realm. The signing of the FTA with Colombia occurred in a moment when internal conflict intensified in Indigenous-occupied areas, usually located at the centre of Canadian oil exploration. Above all, it occurred when denunciations regarding the killing of trade unionists and military and police extrajudicial killings – euphemistically named "false positives" by the Uribe government's officials – were added to Colombia's already poor human rights record (see Engler 2009, 103). This record was undermined not only by military and police members, with the collaboration of the emerging right-wing paramilitary, but also by left-wing guerrilla groups. By considering solely economic-security objectives, the Harper government missed the opportunity to promote Canada's human rights agenda, as well as to strengthen its broad, liberal engagement in the region (188–9).

Concluding Remarks: Greater Engagement with Latin America or Another Split Personality towards the Region under the Trudeau Government?

The election of a Liberal Party Prime Minister, Justin Trudeau, in October 2015, after more than a decade of the Harper Conservatives in office constitutes an opportunity to examine the current and future ascendancy of Canada's split personality thesis as mentioned, as well as its impact on CFSP towards Latin America. At first glance, it seems that the Trudeau government embraces a broad engagement with the world, including Latin America and the Caribbean, and accordingly may pave the way for the harmonization of liberal values and strategic interests. An explanation of values-oriented and strategic components of this government's foreign and security policies could be insightful regarding such a matter.

As for the values-oriented component, such an engagement has entailed embracing a similarly broad, liberal agenda as the Harper government. The Trudeau government continues to consider some security threats which are critical for Latin Americans, such as terrorism; violent conflicts; transnational crime; and the trafficking of drugs, people, and weapons and have, like the Harper government, undertaken some significant strategies to address these issues. Furthermore, its policies towards Latin America and the Caribbean did not reject the objectives of security, prosperity, and democratic governance laid out by its predecessor (GAC 2017, 57).

Above all, like the Harper Conservatives, the Trudeau Liberals have campaigned against regimes turning into dictatorships in the region. The most apparent case has been their stance on the Venezuelan regime of Nicolás Maduro. After all, during the Maduro years, the international community and domestic opposition have witnessed the growing decline and breakdown of democracy and the legal and constitutional order in Venezuela. It has brought about an economic and political crisis that has turned into a humanitarian one, with 5.9 million Venezuelans leaving their country to avoid violence and insecurity and meet their basic needs, including food and medicine (see UNHCR 2021).

At the multilateral level, the Trudeau government has responded to this crisis by passing and supporting several resolutions and statements within the OAS. Above all, this government became an active member of the Lima Group in August 2017 to support "the urgent restoration of democracy and the rule of law in Venezuela" (LG 2020d; see also GC 2020a). The Lima Group was initially founded by Canada, along with Argentina, Brazil, Chile, Colombia, Costa Rica, Guatemala,

Honduras, Mexico, Panama, and Peru. Guyana, Saint Lucia, Bolivia, and Haiti would join later. Under the umbrella of the Lima Group, Canada has denounced the lack of free parliamentary and presidential elections, the corruption and illegal designation of the National Electoral Council, systematic human rights violations, and violence inflicted by Maduro's paramilitary groups against Juan Guaidó, the President of the National Assembly of Venezuela. Indeed, along with several countries, Canada recognized Guaidó as the legitimate interim president of Venezuela after the fraudulent presidential election in 2018 (LG 2020a, 2020b, 2020c, 2020d). Canada and its Latin partners also urged the Office of the Prosecutor of the International Criminal Court to investigate crimes against humanity in Venezuela (LG 2019).

At the unilateral level, the Trudeau government suspended operations at the embassy in Venezuela on 2 June 2019, imposed several targeted sanctions against the Maduro regime's high-level officials, and suspended military cooperation with this country (GC 2020a; LG 2019; CTVnews 2019). Based on its liberal, broad agenda, this government has allocated since 2017 about "$55 million in humanitarian, stabilization and development programming to help respond to the effects of the Venezuela crisis [in the region]" (GC 2020a).

Still, unlike the Harper Conservatives, the Trudeau Liberals have attempted to go further with liberal democracy promotion and restoration. Rather than solely embracing the universal values of freedom, democracy, and human rights which would probably mean to adopt a sort of sanctimonious idealism, the Trudeau government has held the values of "inclusion, compassion, accountable governance, and respect for diversity and human rights" (DND 2017, 61). It has also gone hand in hand with a feminist international assistance policy aimed to "support ... gender equality and the empowerment of women and girls [as] the best way to build a more peaceful, more inclusive and more prosperous world" (GAC 2017, iv). This means that gender equality promotion has ended up being instrumental in advancing the Trudeau government's foreign and security policies.

Concerning the strategic component, security and prosperity remain critical Canadian interests as they constitute, according to this government, the backbone of the Canadian way of life. Since these interests depend on global stability, the rules-based international order and the principle of collective defence, the Trudeau government has made a long-term commitment across the full spectrum of military and peace operations (DND 2017, 59). Rather than embracing the need for committing with the strengthening of Canadian Forces' ability for military intervention exclusively and, consequently, brushing aside peace-keeping operations and soft power strategies – as the Harper government did in several cases – the Trudeau government has included within its action plan "from humanitarian assistance and disaster relief, to peacekeeping, to combat" (DND 2017, 11). It follows that rather than privileging NATO military operations over UN peace missions, Canada has sought to combine both operations during this government.

On the surface, such values-oriented and strategic components seem have been combined by embracing pragmatic idealism, or liberal realism. As Chrystia Freeland the Minister of Foreign Affairs between 2017 and 2019 pointed out, "our objective is to restore constructive Canadian leadership in the world and to promote Canada's interests and values; serve our security and economic interests; and contribute to a more peaceful and prosperous world. To do so, we must build a closer link between our foreign, defence, humanitarian, development, and trade policies. Combined with our upcoming International Assistance Policy and progressive, feminist foreign policy, [our] defence policy will help accomplish our shared objectives" (GAC 2017, 7).

These values and interests informing the Trudeau government's conduct of operations at the global level are critical at the hemispheric level as well (see DND 2017, 61). At first sight, they can help dismantle unequal and violent structures of power dominating some segments of Latin American societies (see Hilgers and Macdonald 2017, 10). With foreign and security policies aimed at combining the values of accountable governance and human rights with those of inclusion, and respect for diversity, within a feminist and comprehensive approach to security, Canada can help mitigate centuries of exploitation and exclusion in the region (see 11).

More importantly, such involvement could bring about a harmonious combination between interests and values. Nevertheless, a closer analysis of the Trudeau government's approach to this region reveals similar flaws as those observed with the Americas Strategy. Firstly, this government does not have a concrete official policy statement similar to the Americas Strategy, which casts its involvement in the region into doubt. After all, to make progress from words to deeds, resource allocations derived from a specific policy framework are indispensable.

Secondly, the rebranding of Canadian foreign policy, mainly the combination of liberal internationalism with a comprehensive and feminist approach to security, has not yet resulted in a real engagement with the region beyond some specific cases. Among these can be highlighted the Canadian support of the Colombian peace process by endorsing local organizations working on humanitarian mine clearance and incorporating female mine clearers. For Colombians, this initiative is significant, as landmines have killed over 12,400 people since 1990, and women, with some exceptions, have not been seen as agents involved in peace and security efforts (GAC 2017, 57 and 63). However, such initiatives are not sufficient to underpin a grand strategy towards the region.

Thirdly, the Trudeau government's involvement in the region has not seemed to have shifted away from the previous government's economic security agenda. Specifically, this agenda has had an impact on its stance on migration, the world drug problem, and the support of Canadian companies, which had, in turn, undermined central aspects of its liberal foreign and security policies towards the region. Regarding migration issues, the Trudeau government has been uninterested in revising the Safe Third Country Agreement (STCA) with the United States. This has occurred even though the Canadian Council for Refugees, Amnesty International, and the Canadian Council of Churches have challenged the STCA before the Federal Court due to the implicit violation of Mexican and Central American refugees' rights (Federal Court, 2020). Instead, the Trudeau government has remained in support of the agreement to avoid bothering its neighbour. After all, it could have compromised its position during the talks that replaced NAFTA by a new Canada–United States–Mexico Agreement (CUSMA), signed in November 2018 and entered into force on 1 July 2020 (see GC 2020b).

The STCA, which was signed for the United States and Canada on 5 December 2002, and came into effect in 2004, was conceived as part of the US–Canada Smart Border Action Plan to control the flow of refugees within North America in the aftermath of 9/11. Not only does it stipulate that refugees must claim asylum in the country of their first arrival, it also considers both Canada and the United States as safe countries for refugees (see GC 2020c). In practical terms, it means refugees cannot make claims from the United States to enter Canada. This restriction has been particularly detrimental for refugees' interests from Mexico and countries in Central America's northern triangle – Guatemala, Honduras, El Salvador – arriving in the United States during the Trump years. Mexican and Central American refugees come to the United States to escape the highest rates of lethal violence globally, including femicide and child abuse. However, they have had to deal with the Trump government's "zero tolerance" policy that has made the United States unsafe for them and, accordingly, has undermined the STCA's underlying rationale (Smith and Hoffman, 2019; Anzueto, 2018; Heilman, 2018). After all, this policy criminalizes asylum seekers at the border, rejects gender- and gang-violence-based asylum claims, conducts illegal detentions, entailing the separation of parents and their children, and grants refugees erratic access to courts (Smith and Hoffman, 2019).

Despite this apparent situation, the Trudeau government has prevented itself from revising the STCA, which would have been required to absorb a higher number of Mexican and Central American refugees and, therefore, mitigate the humanitarian crisis in Canada's backyard (Korducki 2019). With the North American economic-security agenda in mind, the Trudeau government has squandered the opportunity of implementing a key component of its officially stated liberal engagement towards the region. This decision is even more surprising considering that Canada has become one of the countries resettling more refugees in the world during the Trudeau government. Without revising the STCA, the privileged regions have been mostly from the Middle East, Africa, and Asia (see Keung 2019; see also Korducki 2019).

The Canadian stance on the drug problem has been another issue within its foreign and security policy towards the region impacted by North American economic security during the Trudeau government. Indeed, the NAFTA talks could have been one of the primary reasons this government adhered to the Global Call to Action on the World Drug Problem, led by the Trump government. In adhering to this call, the Trudeau government has undermined its liberal engagement with the region. Such an engagement would have entailed not only rejecting Trump's approach but also echoing alternatives in the region to the US historically held approach to this issue. Since 2013, such alternatives have proposed to legalize and decriminalize drug use, replace criminal justice by health-oriented interventions, focus on poverty-alleviation policies or promote a human rights perspective. The underlying rationale of this change promotion relies on the economic and social unintended consequences derived from the conventional approach, especially in the form of increased levels of violence. After all, the conventional approach has entailed a direct fight against drug cartels, gangs, guerrillas, paramilitary groups and other criminal networks taking advantage of the illegal market of drugs. What is more, such a fight has reaffirmed the pervasive structures of violence in Latin America without achieving its primary goal: "a world free of illicit use of drugs" (HRW, 2016). The Trudeau government's decision to adhere to the new version of the War on Drugs is even more contradictory, considering that it adopted a different approach at the domestic level by legalizing cannabis for recreational purposes on 17 October 2018 (see Nolen 2018).

Finally, the economic-security approach, as being promoted by the previous governments in the region by supporting Canadian mining companies, seem to have been equally critical during the Trudeau government. Even though the Liberal government has encouraged Canadian companies to "reflect Canadian values, respect human rights and operate responsibly," there are no details about how specific measures can be undertaken to fulfil such a goal. As Stephen Brown and Liam Swiss point out, "such measures could involve the creation of binding accountability mechanisms that would allow Canadian companies and their foreign subsidiaries to be sued in Canadian courts for acts committed in developing countries, where judicial remedies may be harder to obtain" (Brown and Swiss 2017, 125–6). Without such measures, the Trudeau government will hardly harmonize its liberal values and strategic interests in the region.

To conclude, it can be argued that without a concrete official policy statement backed by specific resource allocations aimed at pinpointing how feminist and comprehensive approaches to security can be concretized in Latin America and the Caribbean, the already existent economic security interests in North America and the rest of the American hemisphere will end up being dominant once again. The result will be a contradictory policy towards Latin America and the Caribbean unwilling and unable to combine liberal values and strategic interests. The Canadian government's ability to undertake a more assertive and coherent engagement regarding security and defence issues will depend on articulating these components and, accordingly, not brushing aside any of them. An exclusively liberal engagement tends to bring about poor calculation of bureaucratic and economic resources and, accordingly, to be synonymous with hollow words, whereas a narrow realist approach to the region might undermine Canada's image since Latin Americans might see its goals, as Peter McKenna (2018, 26) pointed out, as "one-sided," "short-sighted," or even "exploitative."

NOTES

1 The first defence attaché, stationed in Mexico City in 1991, was responsible not only for its host country but also for Belize, El Salvador, Guatemala, Honduras, Nicaragua, and Panama. Since then, Canada has opened four other defence attachés in Brazil (responsible also for Guyana and Suriname), in Chile (responsible also for Argentina, Paraguay, and Uruguay), in Colombia (responsible also for Bolivia, Ecuador, Peru, and Venezuela), and in Cuba (responsible for the Dominican Republic, Jamaica, and Trinidad and Tobago (DND 2014).

2 On the evolution of Canadian Official Development Assistance (ODA) to Latin America and the Caribbean as compared to other regions, as well as

the mixed motives informing ODA towards the region with special regard to three emblematic cases (Peru, Honduras, and Haiti), see Macdonald and Ruckert (2016).

REFERENCES

Anzueto, Marc-André. 2018. "Canada's Approach toward Central American Migration." *Policy Options* (10 August). Accessed 25 October 2021. https://policyoptions.irpp.org/magazines/august-2018/canadas-approach-toward-central-american-migrants/.

Baranyi, Stephen. 2014. "Canada and the Security-Development Nexus in Haiti: the "Dark Side" or Changing Shades of Gray?" *Canadian Foreign Policy Journal* 20, no. 2: 163–75.

Benítez, Raúl, and Athanasios Hristoulas. 2012. "Mexico and Canada: Confronting Organized Crime through Enhanced Security Co-operation." In *Canada Among Nations, 2011–2012: Canada and Mexico's Unfinished Agenda*, edited by Alex Bugaliskis and Andrés Rozental, 221–32. Montreal and Kingston: McGill-Queen's University Press.

Bosold, David, and Wilfred Von Bredow. 2006. "Human Security. A Radical or Rhetorical Shift in Canada's Foreign Policy?" *International Journal*, 61, no. 4: 829–44.

Brown, Stephen, and Liam Swiss. 2017. "Canada's Feminist International Assistance Policy: Game Changer or Fig Leaf?" In *How Ottawa Spends 2017–2018 @150*, edited by Katherine Grahan and Allan M. Maslove, 118–32. Ottawa: School of Public Policy and Administration – Carleton University.

Burtt, Michael. 2009. "Tighter Border Security and Its Effects on Canadian Exports." *Canadian Public Policy/Analyze de Politiques* 35, no. 2: 149–69.

Buzan, Barry. 1991. "New Patterns of Global Security in the Twenty-First Century." *International Affairs (Royal Institute of International Affairs 1944-)* 67, no. 3: 431–51.

Canadian International Development Platform (CIDP). 2020. "Canada's Foreign Aid." Accessed 25 October 2021. https://cidpnsi.ca/anada-foreign-aid-2012-2/.

Clarkson, Stephen, and Maria Banda. 2003. "Paradigm Shift or Paradign Twist? The Impact of the Bush Doctrine on Canada." Paper presented at the *Conference Canada, Free Trade and Deep Integration in North America: Revitalizing Democracy, Upholding the Public Good*, 1–16. Toronto: Centre for Research on Latin America and the Caribbean (CERLAC) at York University and the Canadian Centre for Policy Alternatives (CCPA).

CTV News. 2019. "Canada Suspends Operations at Embassy in Venezuela." Accessed 25 October 2021. https://www.ctvnews.ca/politics/anada-suspends-operations-at-embassy-in-venezuela-1.4448532.

Eayrs, James. 1966. *Right and Wrong in Foreign Policy*. Toronto: Toronto University Press.

Engler, Ives. 2009. *The Black Book of Canadian Foreign Policy*. Vancouver: Red Publishing.

Gotlieb, Alan. 2004. *Romanticism and Realism in Canada's Foreign Policy*. Toronto: C.D. Hove Institute.

Hampson, Fen, and Dean Oliver. 1998. "Pulpit Diplomacy: A Critical Assessment of the Axworthy Doctrine." *International Journal* 20, no. 3: 305–19.

Heilman, Jayme. 2018. "Canada is Turning Its Back on Central American Refugees." *Policy Options* (20 June). Accessed 25 October 2021. https://policyoptions.irpp.org/magazines/ana-2018/anada-is-turning-its-back-on-central-american-refugees/.

Hilgers, Tina, and Laura Macdonald. 2017. "How Violence Varies: Subnational Place, Identity and Embeddedness." In *Violence in Latin America and the Caribbean: Subnational Structures, Institutions and Clientelistic Networks*, edited by Tina Hilgers and Laura Macdonald, 1–38. Cambridge: Cambridge University Press.

Human Right Watch (HRW). 2016. "Rethinking the War on Drugs: A Rights Perspective." Accessed 25 October 2021. https://www.hrw.org/blog-feed/rethinking-war-drugs.

Jockel, Joseph T., and Joel J. Sokolsky. 2009. "Canada and NATO: Keeping Ottawa in, Expenses Down, Criticism Out ... and the Country Secure." *International Journal* 64, no. 2: 315–36.

Keung, Nicholas. 2019. "Canada Urged to 'Step Up and Help' More Central American Refugees." *The Star* (20 August). Accessed 25 October 2021. https://www.thestar.com/news/anada/2019/08/20/anada-urged-to-welcome-more-central-american-refugees.html.

Klepak, Hal. 2012. "The Most Challenging of Links? Canada and Inter-American Security." In *Canada Looks South: In Search of an Americas Policy*, edited by Peter McKenna, 27–57. Toronto: University of Toronto Press.

Korducki, Kelli María. 2019. "A Blind Spot for Central America in Canada's Refugee Response." *Opencanada*. https://opencanada.org/blind-spot-central-america-canadas-refugee-response/.

Macdonald, Laura. 2016. "Canada in the Posthegemonic Hemisphere: Evaluating the Harper Government's Americas Strategy." *Studies in Political Economy* 97, no. 1: 1–17.

Macdonald, Laura, and Arne Ruckert. 2016. "Continental Shift? Rethinking Canadian Aid to the Americas." In *Rethinking Canadian Aid*, edited by Stephen Brown, Molly Den Heyer, and David Black, 133–50. Ottawa: University of Ottawa Press.

Mace, Gordon, Jean-Philippe Thérien, and Stefan Gagné. 2012. "Canada and the Security of the Americas. Between Old Threats and New Challenges." *International Journal* 67, no. 3: 603–22.

McKenna, Peter. 1994. *Canada and the OAS*. Ottawa: Carleton University Press.

– 2012. "Preface." In *Canada Looks South: In Search of an Americas Policy*, edited by Peter McKenna. Toronto: University of Toronto Press.

– 2018. "Canada and Latin America: 150 Years Later." *Canadian Foreign Policy Journal* 24, no. 1: 18–38.

Melakopides, Costas. 1998. *Pragmatic Idealism. Canadian Foreign Policy, 1945–1995*. Montreal and Kingston: McGill-Queen's University Press.

Miller, Eric. 2016. *Engagement and Pragmatism: Towards an Enduring Canadian Strategy in Latin America*. Calgary: Canadian Global Affairs Institute.

Moens, Alexander. 2007. "Afghanistan and the Revolution in Canadian Foreign Policy." *International Journal* 63, no. 3: 569–86.

Morden, Reid. 2012. "Hemispheric Security: The Canada-Mexico Conundrum." In *Canada among Nations, 2011–2012: Canada and Mexico's Unfinished Agenda*, edited by Alex Bugaliskis and Andrés Rozental, 214–20. Montreal and Kingston: McGill-Queen's University Press.

Nolen, Stephanie. 2018. "Canada Signs on to U.S.-Led Renewal of War on Drugs." *The Globe and Mail* (24 September). Accessed 25 October 2021. https://www.theglobeandmail.com/world/article-canada-sign-on-to -us-led-renewal-of-war-on-drugs/.

Nossal, Kim. 1999. Review [*The Pragmatic Idealism: Canadian Foreign Policy* by Costas Melakopides]. *Canadian Journal of Political Science/Review Canadienne de Science Politique* 32, no. 4: 786–7.

– 2007. "Right and Wrong in Foreign Policy 40 Years on. Realism and Idealism in Canadian Foreign Policy." *International Journal* 62, no. 2: 263–77.

Organization of American States (OAS). 2003. *Special Conference on Security – Declaration on Security in the Americas* (27–8 October). Mexico City, Mexico.

Rochlin, James. 1994. *Discovering the Americas. The Evolution of Canadian Foreign Policy Towards Latin America*. Vancouver: UBC Press.

Rodríguez, Federmán. 2017. *Canadá y la Seguridad Internacional. Discursos Liberales y Realistas de una Potencia Media en Transformación*. Bogotá: Centro Editorial Universidad del Rosario.

Sands, Christopher. 2008. "An Independent Security Policy for Canada in the Age of Sacred Terror." In *An Independent Foreign Policy for Canada? Challenges and Choices for Future*, edited by Brian Bow and Patrick Lennox, 103–17. Toronto: University of Toronto Press.

Shamsie, Yasmine, and Ricardo Grinspun. 2010. "Missed Opportunity: Canada's Re-Engagement with Latin America and the Caribbean." *Canadian Journal of Latin American and Caribbean Studies* 35, no. 69: 171–99.

Smith, Craig Damian, and Stephanie Hofmann. 2019. "Will Canada Suspend Its Safe Third Country Agreement with the United States?" *Foreign Policy* (6 November). Accessed 25 October 2021. https://foreignpolicy.com/2019 /11/06/anada-suspend-safe-third-country-immigration-united-states/.

Sokolsky, Joel, and Philippe Lagassé. 2005/2006. "Suspenders and a Belt: Perimeter and Border in Canada-U.S. Relations." *Canadian Foreign Policy* 12, no. 3: 15–29.

Stevenson, Brian J.R. 2000. *Canada, Latin America, and the New Internationalism. A Foreign Policy Analysis 1968–1990*. Montreal and Kingston: McGill-Queen's University Press.

Tomlin, Brian, Norman Hillmer, and Fen Osler Hampson. 2008. *Canada's International Policies. Agendas, Alternatives and Politics*. Oxford: Oxford University Press.

Walby, Kelvin, and Jeffery Monaghan. 2011. "'Haitian Paradox' or Dark Side of the Security-Development Nexus? Canada's Role in the Securitization of Haiti, 2004–2009." *Alternatives: Global, Local, Political* 36, no. 4: 273–87.

Official Documents

Department of Foreign Affairs and International Trade (DFAIT). 1995. *Canada in the World. Canadian Foreign Policy Review 1995*. Accessed 8 December 2021. https://www.dfait-maeci.gc.ca/foreign_policy/cnd-world/menu-en.asp.

Department of National Defence (DND). 1994. *1994 White Paper on Defence*. Accessed 8 December 2021. https://publications.gc.ca/collections/collection_2012/dn-nd/D3-6-1994-eng.pdf.

– 2008. *Canada First Defence Strategy*. Ottawa, Ontario.

– 2014. *The Canadian Armed Forces' Engagement in the Americas*. Accessed 8 December 2021.http://www.forces.gc.ca/en/news/article.page?doc=the-canadian-armed-forces-engagement-in-the-americas%2Fhie8w7sf.

– 2017. *Strong, Secure, Engaged. Canada's Defence Policy*. Ottawa, Ontario.

Federal Court. 2020. *Canadian Council for Refugees v. Canada (Immigration, Refugees and Citizenship)*, 2020 FC 770. File numbers IMM-2229–17, IMM-2977–17, IMM-775–17, 20 July 2020.

Global Affairs Canada (GAC). 2017. *Canada's Feminist International Assistance Policy*. Ottawa, Ontario.

Government of Canada (GC). 2005a. *Canada's International Policy Statement: A Role of Pride and Influence*. Ottawa, Ontario.

– 2005b. *Canada's International Policy Statement: A Role of Pride and Influence - Overview*. Ottawa, Ontario.

– 2005b. *Canada's International Policy Statement: A Role of Pride and Influence - Diplomacy*. Ottawa, Ontario.

– 2005c. *Canada's International Policy Statement: A Role of Pride and Influence - Defence*. Ottawa, Ontario.

– 2005d. *Canada's International Policy Statement: A Role of Pride and Influence - Development*. Ottawa, Ontario.

– 2009. *Canada and the Americas. Priorities & Progress.* Ottawa, Ontario.
– 2013. *Building Resilience against Terrorism: Canada's Counter-Terrorism Strategy.* Ottawa, Ontario.
– 2016. *Evaluation of the Stabilization and Reconstruction Task Force (START) and Global Peace and Security Fund (GPSF).* Accessed 25 October 2021. https://www.international.gc.ca/gac-amc/publications/evaluation/2018/start-gpsf.aspx?lang=eng.
– 2018. Embassy of Canada in Haiti. "Canada-Haiti Relations." https://www.canadainternational.gc.ca/haiti/bilateral_relations_bilaterales/canada_haiti.aspx?lang=eng.
– 2020a. "Canada and the Venezuela Crisis." Accessed 25 October 2021. https://www.international.gc.ca/world-monde/issues_development-enjeux_developpement/response_conflict-reponse_conflits/crisis-crises/venezuela.aspx?lang=eng.
– 2020b. "A New Canada-United States-Mexico Agreement." https://www.international.gc.ca/trade-commerce/trade-agreements-accords-commerciaux/agr-acc/cusma-aceum/index.aspx?lang=eng.
– 2020c. "Canada-U.S. Safe Third Country Agreement." Accessed 25 October 2021. https://www.canada.ca/en/immigration-refugees-citizenship/corporate/mandate/policies-operational-instructions-agreements/agreements/safe-third-country-agreement.html.
Government of Colombia. 2003. *Política de Defensa y Seguridad Democrática.* Bogotá, Colombia.
Lima Group (LG). 2019. Lima Group Declaration. Available from: www.international.gc.ca [4 January 2020].
– 2020a. "Lima Group Ministerial Meeting February 20, 2020 Chair's Statement." Accessed 25 October 2021. https://www.canada.ca/en/global-affairs/news/2020/02/lima-group-ministerial-meeting-february-20-2020-chairs-statement.html.
– 2020b. "Lima Group Statement." (2 March). Accessed 25 October 2021. https://www.international.gc.ca/world-monde/international_relations-relations_internationales/latin_america-amerique_latine/2020-03-02-lima_group-groupe_lima.aspx?lang=eng.
– 2020c. "Lima Group Statement." (2 April). Accessed 25 October 2021. https://www.international.gc.ca/world-monde/international_relations-relations_internationales/latin_america-amerique_latine/2020-04-02-lima_group-groupe_lima.aspx?lang=eng.
– 2020d. "Lima Group Statement." (16 June). Accessed 25 October 2021. https://www.international.gc.ca/world-monde/international_relations-relations_internationales/latin_america-amerique_latine/2020-06-16-lima_group-groupe_lima.aspx?lang=eng&_ga=2.160750013.1818603230.1635182040-80533344.1635182040.

National Defence Department (DND). 2017. *Strong, Secure, Engaged. Canada's Defence Policy.* Ottawa, Ontario.

Organization of American States (OAS). 2016. *The Mission to Support the Peace Process in Colombia (MAPP/OAS).* Accessed 25 October 2021. https://www.oas.org/en/media_center/press_release.asp?sCodigo=S-017/16.

United Nations High Commissioner for Refugees (UNHCR). 2021. "Venezuelan Situation." https://www.unhcr.org/venezuela-emergency.html.

5 Canada and Democracy Promotion: The 2015 Electoral Crisis in Haiti

YASMINE SHAMSIE

Canada has been helping foster democracy in the Americas for almost three decades now. It has provided aid to elections, the penal and justice sectors, legislatures, independent media, and civil society.[1] While it has deployed democracy-building initiatives across the hemisphere, Haiti stands out. Not only has the country been a priority engagement for Canada's foreign policy, but it has received consistent and wide-ranging democracy aid since 1991. Among its democracy-reinforcing objectives, Haiti's elections have been a primary and sustained focus for Ottawa. This chapter considers Canada's reactions, conduct, and bearing during Haiti's 2015 electoral crisis to help us better understand its approach to democracy-building in the hemisphere and the implications of this approach for the countries it assists.

In October 2015, Haiti's presidential election went awry, producing a year-long political crisis and a contentious election redo. I argue that the model of democracy-building Ottawa has employed in Haiti explains its diplomatic stance during the crisis and, more importantly, the fact it missed an opportunity to encourage and support a democratic growth spurt. I also suggest that going forward, Canada should restrict its support to Haiti's elections to the most technical of tasks. This admittedly controversial recommendation stems from the fact that, in my view, Canada has become overly enmeshed in Haiti's elections, what should be a domestic, sovereign affair, with significant implications for Haiti's democratic development.

Canadian Democracy Promotion Efforts in Haiti

The relationship between Canada and Haiti has been solid and lasting. As Stephen Brown observes, "Haiti is a natural fit for Canadian aid, as it is a low-income country in Canada's neighborhood and one with which

Canada shares an official language, otherwise uncommon in the Americas, and a long history of development cooperation" (Brown 2018, 154). This enduring relationship has produced a Montreal-based diaspora numbering more than 100,000, which has at times influenced Canadian engagement on Haiti (Patriquin 2006). For instance, concerns from the diaspora contributed to Ottawa's decision to support the 1994 US intervention that restored President Aristide to power (Keating 2001, 220).

Canadian aid to Haiti began in 1968 but intensified in the 1980s following the fall of the Duvalier dictatorships. Since that time, Canada's objectives have been to reduce poverty, strengthen security, and foster democracy (CIDA 2004, 5; Shamsie 2006; Baranyi 2007; Muggah 2007). From 2004 onward, Ottawa has regularly provided about $100 million a year in aid to Haiti, which makes it the second most important bilateral aid donor after the United States (Baranyi 2010). In response to the country's 2010 devastating earthquake, individual Canadians pledged $220 million, which was matched by the federal government. Since 2004, when the United Nations Stabilization Mission in Haiti (MINUSTAH) arrived, Haiti has been Canada's top aid recipient in the Americas (Global Affairs Canada 2017).

Assistance to democracy has been one of the most controversial aspects of Canadian aid to Haiti.[2] This is not due to its funding priorities given it supports the same activities as most Western governments: elections, institutions, media, civil society actors. It is Ottawa's diplomatic efforts that have raised concerns. For instance, a number of scholars and activists have condemned Canada's actions before and during the 2004 political crisis that forced President Aristide into exile (Fenton 2004; Engler and Fenton 2005; Hallward 2007; Sanders 2007). They believe that one year before Aristide's forced departure, Canada conspired with the United States and France to have him removed. Whether an actual plan was devised is difficult to know. However, it is clear that when President Aristide asked the international community for help, Canada did not respond.[3] If regime change was not the objective, it is unclear why Canada and other international actors stood by and tacitly supported a band of insurgents, led by former army and death squad members, against the country's elected government. This diplomatic episode forms part of Canada's democracy-building portfolio. It shapes how Haitians view subsequent democracy-promoting activities. In short, Canada's democracy aid, including our support to elections, is viewed, interpreted, and judged alongside these interventions.

Whether countries explicitly state it or not, they have a democracy-building approach which is underpinned by a set of assumptions about what democracy is and how it develops. Elsewhere, I have argued that Canada employs a democracy-building model that Thomas Carothers (1999, 91) has dubbed the "natural transition sequence combined with institutional modelling" approach (Shamsie 2008). It posits that democracy emerges through a series of progressive stages: holding of elections, which lead to a legitimate elected government, culminating in a gradual democratic consolidation. A major assumption of this approach is that replicating the institutions of already existing Western democracies is key to this process (Carothers 1999, 90). This explains why Canada's democracy assistance involves a good deal of training and capacity-building. Its sustained aid to elections, police, judicial reform, and the penal system in Haiti is evidence of this approach. In the past, it has also provided training to Haitian election observers and supported the training of political candidates and journalists (Baranyi 2008). Canada has also helped to develop the Haitian parliament's ability to produce legislation and improve its relations and ability to dialogue with the Executive (Baranyi 2007).

Another belief that underpins Canada's democracy-building approach is that democratization is a relatively elite enterprise that should ideally unfold in an incremental, managed, and orderly fashion. While civil society engagement is encouraged, political and economic elites are considered the main crafters of democratic transitions, with civil society participating through designated institutional channels, rather than in direct, unmediated ways. Indeed, prominent democratization scholars have noted the value of limited and modest participation, particularly during the transition. Larry Diamond (1993, 103), for instance, has argued that "Democracy implies dissent and division, but on the basis of consent and cohesion. It requires that the citizens assert themselves, but also that they accept the government's authority. It demands that citizens care about politics, but not too much."

A third important aspect of Canada's democracy aid to Latin America, and Haiti in particular, has been its deep connection to the economic project of neoliberalism. Canada's democracy-promotion efforts have been rolled out in the context of neoliberal market reforms and the opening up of the hemisphere to free trade. Indeed, structural adjustment has been a key aspect of the economic development plans devised for Haiti by Canada and other international donors since 1990 (Dupuy 1994; 2007). Canada's democracy aid, therefore, has been shaped to be consistent with its support for an export-led development strategy, and the economic prescriptions of the World Bank and the course of IMF-imposed restructuring (Burron 2011). This economic stance reinforces Ottawa's predisposition to order and measured political change given these global agencies recommend attracting foreign investment to spur economic growth, and investors demand a stable and predictable political environment. Because Canada views order and stability, pre-requisites for capitalist growth, as essential to both democratic and economic development it is not surprising that the form of democracy it has supported would value both.

Key to democratization, for Canada, are the behaviour of political actors and the formal structures of the political system. Aid is, therefore, directed at the institutional, technical, and procedural aspects of democracy. There is little attention given to the hidden structures of power that exist in societies like Haiti or to the notion that political engagement is blocked by poverty and privilege. Because the form of democracy Canada promotes, often referred to as low-intensity democracy, places little emphasis on socio-economic inequalities, disrupting pre-democracy economic and social arrangements is not envisioned. As such, democracy is viewed as disconnected from notions of social justice (Robinson 1996). The democratization literature acknowledges this tendency, noting that what leads hesitant political actors tied to the previous regime to accept democracy "is that in the inaugural phase, coexistence usually takes precedence over any radical social and economic programs" (Di Palma 1990, 22–4).

Finally, it is important to note that post-9/11, Canada's efforts to support democracy have occurred within a broader security framework aimed at maintaining order and stability in so-called fragile states[4] (Engler and Fenton 2005; Cameron and Hecht 2008; Shamsie 2008; Burron 2011). Since 9/11, Canada, has made establishing order in "chaotic" environments a foreign-policy objective. Indeed, its former aid agency, the Canadian International Development Agency (CIDA), was given the additional mandate of supporting "international efforts to reduce threats to Canadian security" (Simpson and Tomlinson 2006). Soon after, Canada piloted the OECD-DAC Principles for engaging with fragile states in Haiti (CIDA 2007, 37). Hence, from 2004 (the start of the MINUSTAH), Canada has promoted democracy while simultaneously supporting a peace enforcement or pro-stability agenda (Shamsie 2009; Baranyi 2012; Burron and Silvius 2013). That desire to encourage order and maintain stability during the democratic transition has once again been reinforced in Canadian foreign-policy priorities.

The First Round: The 9 August Vote

At the heart of the turmoil that led to the redo of Haiti's 2015 presidential election are different accounts of election day – specifically the electoral irregularities that took place. The significance and the degree of fraud reported were not accorded the same level of importance by the various observer groups. The reasons for this discrepancy need to be explored more fully, but a first round of interviews and early accounts indicate that, for international observers, redoing an election would be justified only if the irregularities were so egregious so as to change the final result. Local observer groups appeared to give more weight to the gravity of the irregularities they encountered. They also felt that the *possibility* of altered the final results – rather than certainty that the results would be different – could warrant a redo.

The conditions for an electoral crisis were ripe in 2015 because Haiti was scheduled to run a large number of elections. For instance, by January 2015, the terms of 119 House of Deputies members had expired, as had the terms of 20 out of 30 senators. At the local level as well, elections were overdue. As the terms of mayors and municipal deputies lapsed, President Martelly had appointed interim replacements. Because of so many overdue elections, the legislative branch was no longer functioning, and Martelly began ruling by decree in January 2015.

Haiti's electoral system is a two-round system. If the top candidate does not have 50 per cent of the vote or a 25 per cent lead over the nearest competitor, a second round of voting is required. On 9 August, elections were held to fill seats in the House of Deputies and Senate, as well as local municipal posts. These elections were quite chaotic and marked by numerous irregularities. At the same time, observers expressed different levels of confidence in the process. One Haitian observation group, the Citizen Observatory for Institutionalizing Democracy (OCID), felt the elections were "more or less acceptable."[5] The European Union/OAS observer mission agreed, although it expressed concerns about incidents of violence at 202 of the 1,508 voting centres (OAS 2017, 3). The EU/OAS report also noted that political party members charged with overseeing the process at polling stations (known as "*mandataires*") did not receive their credentials on time, leading to credential cards being issued at the polls on the day of voting. As we shall see below, this practice had momentous implications for voting during the second round.

In contrast to the evaluations of the elections as acceptable if flawed presented by the joint EU/OAS team and the Haitian OCID observers, a second group of Haitian civil society actors was less approving in its assessment of the first round. A joint Haitian observation group made up of the National Network for the Defense of Human Rights (*Réseau national de Défense des Droits Humains* [RNDDH]), the National Observation Council (*Conseil national d'Observation* [CNO]), and the Haitian

Council of Non-State Actors (*Conseil Haïtien des Acteurs Non Etatiques* [CONHANE]) reported higher levels of violence at the polls, noting that at least half of all the voting centres had experienced violence and intimidation. The group went so far as to question the legitimacy of the new parliament given the extent of the violations and problems that its 1,500 observers noted. The group was also concerned about the voter turnout – only 18 per cent of eligible voters (Charles 2015c). Indeed, it recommended that "all actors involved at every level in the electoral process avoid trivializing the facts recorded during this election," and specifically recommended that the Provisional Electoral Council (CEP), the body that organizes and supervises the elections, and acts as arbiter, "be wary of all those who claim that everything went well."[6]

The 25 October Vote: A Debacle ... or Not

The second round of parliamentary elections (Senate and House of Deputies) and the first round of the presidential vote took place on 25 October 2015. The focus of all concerned, however, was on the presidential vote. The results of this first round for president were as follows: first-place finisher, Jovenel Moïse (32.8 per cent) of the ruling *Parti Haïtien Tèt Kale* (PHTK) party, in second place Jude Célestin (25.3 per cent) of the *Ligue Alternative pour le Progrès et L'Émancipation Haïtienne* (LAPEH), an affiliate of former president René Préval. These two men would be in the run off, or second round. The third-place finisher, Moïse Jean-Charles of the *Pitit Dessalines* platform received 14.3 per cent, while Maryse Narcisse of *Fanmi Lavalas*, Jean Bertrand Aristide's party, came in fourth with about 7 per cent.

Serious allegations of fraud emerged almost immediately after the vote. But, once again, there was significant disagreement regarding the levels of fraud encountered and its implications. Straight away, two camps emerged. The first was comprised of most international donors, including Canada, and observers from the OAS, EU, and OCID. The second camp included the National Network for the Defense of Human Rights (RNDDH)[7], a group made up of various churches, and a final group comprised eight candidates who did not win in the presidential vote – known later as the G8.

The irregularities that day were significant but once again, as in the first round, were accorded different degrees of importance by the different camps. For instance, the RNDDH-led observation team, which had deployed observers in 76 per cent of the polling stations, witnessed ballot-box stuffing as well as incidents of political party representatives observing the vote (the *mandataires*) voting multiple times. The effect of *mandataires'* voting severely discredited the results for this group. For others, including the OAS team, the *"mandataire"* issue was viewed as unfortunate and problematic but not significant enough to alter the final results, therefore not more problematic than other irregularities that took place that day. Because the *mandataire* problem played such a prominent part in the ultimate decision to rerun the elections, it is worth describing in some detail.

The CEP had accredited more than 900,000 *mandataires* to observe the voting process in polling stations (OAS 2017, 7). Many viewed the accreditation method – blank cards given to each party by polling station – as a serious problem because it allowed *mandataires* to vote without being on the electoral list and made it impossible to know whether the persons with these accreditation cards were who they said they were (OAS 2017, 7). Moreover, because the cards were distributed prior to election day, they wound up being bought and sold illegally. The buying and selling of cards was witnessed by all electoral observer groups but accorded less weight by the OAS/EU and OCID. This was, once again, because they believed it would not have affected the final results.

When the results were finally published, the second-place finisher, Jude Célestin, Catholic bishops and religious leaders from various Protestant groups, as well as the G8 candidates demanded the establishment of a commission to investigate what they called unacceptable levels of fraud (Charles 2015a). Canada and other international actors did not immediately support calls for a commission, concerned that it would cause the second presidential round to be postponed. Since President Martelly was scheduled to step down, a postponement could lead to the appointment of an interim government – something international donors hoped to avoid at all costs (Guyler Delva 2016).

In response to mounting pressure from second-place finisher Célestin, who refused to advance to the second round without the investigation, and from civil society groups, President Martelly established an *Evaluation Commission* on 22 December 2015. The five-member Commission, appointed by the president and assisted by EU and OAS experts, was given three days to make recommendations (Charles 2015b). G8 candidates and many in civil society felt the restricted mandate was inadequate, given the serious allegations of fraud. Demands for a full investigation, therefore, did not disappear[8] (Charles 2015b).

In the end, the Evaluation Commission found evidence of widespread irregularities, which undermined the legitimacy and integrity of the electoral process for many. Moreover, the Commission confirmed that *mandataires* had voted repeatedly (Johnston 2016). Despite these flaws in the voting process, the Commission did not go so far as to conclude that fraud had taken place. It opted instead for the wording: "irregularities akin to fraud" (Johnston 2016). It did, however, recommend a more thorough verification to determine whether "massive fraud" had indeed taken place, as claimed by opposition candidates and the majority of local election observer groups[9] (Charles 2016).

Initially, President Martelly tried to ignore the Commission's gloomy conclusions, scheduling the second presidential round for late January 2016; however, protests forced the cancellation of that runoff. Notably, even Canada acknowledged that a president elected under these conditions would be lacking in legitimacy. Martelly stepped down, as prescribed by the constitution, leading to the instalment of an interim government, headed by the president of the Senate, Jocelerme Privert. The temporary government was given the mandate to re-establish faith in the electoral process and hold elections in three months.

From an "Evaluation Commission" to an "Independent Electoral Evaluation and Verification Commission"

Privert's interim government faced continuous calls for a deeper inquiry into the 25 October presidential vote. It is interesting that, rather than reassuring Haitians, the report of Martelly's Evaluation Commission had served to drag confidence down to an all-time low. More and more Haitians felt that in order to inject the electoral process with a degree of reliability a more in-depth investigation – independent and Haitian-led – was needed. Privert finally responded on 28 April 2016 by creating a new Independent Electoral Verification and Evaluation Commission (*Commission Indépendante d'Evaluation et Vérification Electorale* [CIEVE]) to review the voting and counting process of the presidential election. It was charged with analysing poll watcher lists, partial electoral lists, results sheets, ballots, and electoral complaints. The new CIEVE was given one month do its work – in comparison to the Evaluation Commission's three days.

Despite the fact that Canada and other international donors expressed reservations, the CIEVE carried out its investigation and delivered its report on 30 May. The report concluded: "the electoral process was marred by serious irregularities, grave inconsistencies, and massive fraud" (Ives 2016).[10] It also stated: "Many acts committed in violation of the law or regulations (including the electoral decree) were systematic (well-organized), and there was intent to deceive (fraud)" (Ives 2016). It recommended the presidential vote be annulled and that the agency charged with organizing elections, the Provisional

Electoral Council (CEP), organize a new election.[11] The CEP complied, announcing a new presidential vote on 9 October 2016, with a second round 8 January 2017.

The CEP's decision to redo the presidential vote prompted strong reactions from international donors. This was not surprising given 200 EU and OAS observers had deemed the original October 2015 vote clean – "by Haitian standards."[12] Canada was frustrated with what it perceived as an interminable electoral process. A CBC News article quoted off-the-record senior government sources, indicating "Canada is fed up with Haiti's leaders playing political games on the donor's dime" (Dyer 2016). Ottawa had expressed its frustration even earlier when the vote was postponed more than once. Upon hearing of the decision to redo the elections, Canada submitted a Diplomatic Note to the OAS Permanent Council requesting a discussion of Haiti's electoral process. The Note expressed concerns that failure to complete the long-delayed electoral process could "affect the constitutional exercise of power," possibly leading to the application of Inter-American Democratic Charter measures to defend Haitian democracy (Noriega 2016, Permanent Mission of Canada at the OAS 2016). The UN, for its part, expressed its displeasure by noting Haiti could "ill afford" a prolonged period of transitional governance, which could "affect international support" (UN News Service 2016). The EU had the strongest reaction, pulling out its electoral observation mission and eventually releasing a very detailed critique of the CIEVE's analysis (Union Europeenne Mission d'Observation Electorale Haïti 2015 Élection Générales 2016). The head of the EU mission characterized CIEVE's report as having "many factual, legal, methodological and conceptual weaknesses" (Euractiv .com and Agence France Press [AFP] 2016). The only supporter of an election redo was the former head of the 2016 OAS special mission to Haiti, Ronald Sanders, who insisted a verification commission was needed to restore faith in the electoral process (Sanders 2016).

The Haitian government stood its ground, however, despite significant external pressure and the expected costs of an election redo. It announced it would commit US$55 million to a new election – most coming from port fees, the Central Bank, and Haiti's National Credit Bank, rather than government programmes (Reuters 2016). Unfortunately, days before the scheduled redo, Hurricane Matthew reached Haiti, causing the worst humanitarian disaster since the 2010 earthquake. Electoral infrastructure was severely damaged in some parts of the country, forcing elections to be rescheduled for 20 November, with a second round on 27 January 2017. In the end, Canada did set aside $1.4 million for the new elections, which supported voting logistics and funded independent observers. On 20 November 2016, Jovenel Moïse won the presidential vote with 55 per cent of the vote, making a second round unnecessary. Some opposition political parties challenged the results, but a special electoral tribunal dismissed their fraud allegations.

The Election Rerun: Why the Resistance?

Canada opposed the rerun for two reasons. First, because it felt organizing a new election would delay rapid action on Haiti's grave problems, a food crisis and a Zika virus outbreak among them. Ottawa was not wrong since the absence of an elected president at the helm can slow the disbursement of international aid. At the same time, urging acceptance of election results amid allegations of fraud and street protests can create the conditions for a serious governance crisis down the road. Given that Canada's overarching objective is to build faith in both democracy and government, rather than resisting the redo, Canada should have given more weight to how Haitians received the CIEVE's recommendation. The CIEVE's findings were supported by a wide swath of Haitian society: media, human rights groups, and, not surprisingly, opposition parties. Put plainly, Canada should have emphasized "process," that is, ensuring the process endowed candidates going into the second round with a solid degree of legitimacy.

The other reason Canada did not favour an election redo was that it trusted the conclusions of OAS/EU observers, who judged the original process acceptable. The problem was that Haitians did not entirely trust the assessment of these outside observers, particularly since there was a clear difference of opinion between international observers and a network of local observer groups. The disagreement stemmed from how many and what types of irregularities invalidate an election. While the OAS/EU observers were not overly troubled by the number of irregularities they documented, local observers were. While more investigation would be needed to explain the difference in perspective here, I believe it stems in part from expectations. The OAS expects to encounter a fairly large number of irregularities, some of them quite egregious. This is reflected in the director of the OAS's Department of Electoral Cooperation and Observation, Gerardo de Icaza's statement after the elections: "No election is perfect – every election has irregularities. What we try to look at is, are those irregularities … a determining factor in the result of the election or not" (Haiti Advocacy Working Group 2016). The OAS focuses on whether the outcome would have been different.

Local observers, perhaps because they are more invested in the process, seem to have higher expectations regarding what is an acceptable level of fraud as well as unacceptable electoral behaviour – which may depart from internationally accepted norms. For instance, for local observers there was a level of outrage associated with the *"mandataire"* issue that was absent from the OAS/EU reports. Locals assigned a profound degree of affront to this particular unlawful behaviour that international observers did not. It is also worth noting that the National Network for the Defence of Human Rights (RNDDH)–led observation team believed the *mandataire* irregularities were so great they did call into question the results.

Finally, the views of local observer groups, regarding the assessments and conclusions of their foreign counterparts, may have been affected by their deep distrust of international actors in general. This general weariness, I believe, has seeped into the electoral arena. For instance, data indicate that in transition countries the public typically trusts the findings of international election observers over domestic observers. Not so in Haiti. According to a study conducted by OCID in July 2015, Haitians have greater faith in their own observers (Observatoire Citoyen pour l'Institutionnalisation de la Démocratie [OCID] 2015, 22–3).

Going Forward: Expecting Electoral Messiness Makes Sense

Canada should expect chaotic elections often accompanied by intense controversies over Haiti's electoral contests for years to come. There are several reasons for this, but two stand out: weak institutions and low levels of social trust. On the matter of institutions, because Haiti's legislature, political parties, and the electoral management body are works in progress, every Haitian election unfolds amid broader processes of political maturation and legitimation. For instance, the legislative branch is regularly slow and late in passing electoral legislation. It has yet to function in an efficient, effective, and representative manner and is regularly the site of intense partisan conflict. These clashes often lead to paralysis, producing delays in the constitutionally determined electoral calendar, which frustrates Canada and other donors.

Failure to respect the electoral calendar can also be attributed to Haiti's political parties, which need greater maturity as well. Few are broad-based with policy agendas. They tend to be formed around a particular individual, with very few connections to citizens. As a result, members often break rank with their parties to pursue their own interests. Without party loyalty, political compromises are difficult to strike, making passing electoral laws an unpredictable, slow, and tortuous affair. Those who study democratic development argue that unfortunately international actors can do very little to alter these dynamics (Carothers 1999, 184).

Lastly, the body that is charged with organizing the country's elections lacks capacity and permanence. The framers of Haiti's 1987 Constitution had mandated the first post-dictatorship elections be organized by a "provisional" electoral council (CEP), to be replaced by a "permanent" council. As of 2018, the CEP is still provisional, which means it must be reconstituted prior to every election. This increases the likelihood of electoral chaos for a number of reasons. Because CEP members are appointed prior to each election, expertise is not accumulated, which leads to repeated errors and low confidence in the electoral process. Also, reconstituting the CEP prior to each election entails selecting new members each time, which inevitably triggers tensions and conflict over its potential politicization. In fact, in 2015, opposition parties accused the executive of stacking the CEP with pro-government members. Because the CEP was perceived as partisan, its actions roused widespread mistrust, further fuelling the electoral crisis.

The second reason to expect Haiti's elections to be characterized by delays, disputes, and unrest relates to the low levels of social trust, particularly in the political arena. Because democracy is only twenty-five years old and slavery, colonialism, and despotism have figured prominently in Haitian history, the reservoirs of social trust across society are very low. When trust is low, the legitimacy of the entire process can be easily undermined by a single break in the electoral chain (Norris 2013, 567). That chain is very long and involves many steps – from drafting electoral legislation and voter registration to counting and tabulating the votes and announcing the results. The likelihood of mistakes and problems occurring at one or more of these stages is very high. Given the context of general mistrust and weak poorly functioning institutions, one would expect losers to claim electoral violations and dispute results. Still, Ottawa should not disregard the calls from losers for closer inspection. Each electoral process deserves careful scrutiny and serious evaluation to discourage future fraud and build trust in the democratic process.

An Alternate Perspective on Haiti's Elections

When Haiti announced it would rerun the 2015 presidential vote, Canada was frustrated, some might even say exasperated. Yet, while Haiti's decision was unexpected, it was not puzzling or baseless given the number of irregularities and, more importantly, the findings and recommendations of its own local electoral observers. I suggest that rather than tending towards pessimism, Canada could have viewed the electoral crisis, and its ultimate resolution, through the lens of institution building. That is, as an opportunity for Haitians (politicians, institutions, local electoral observers, and citizen groups) to work through a political crisis and broker a solution through negotiation.

While Ottawa's immediate goal is to support prompt, timely elections, its long-term aim is to encourage the country's democratic development and to boost citizen respect and confidence in its fledgling political institutions. The 2015 presidential election generated a great deal of political turmoil, but it was ultimately channeled into a conflict-resolution instrument – an Independent Verification Commission. This should have reassured Ottawa. Moreover, the Commission's recommendation – to redo the election – was adopted, which should have also been viewed as a good sign, indicating institutional remedies were triumphing over political strife.

There is another element of this electoral ordeal that Canada should have appreciated. The solution that was ultimately devised and carried out – an independent investigation and a redo – was not developed or orchestrated by outsiders. This was encouraging because, although it can sometimes take quite some time for political actors to devise a solution, the process can contribute to institution building. For instance, prolonged deliberations can help build trust among political actors in a society where it is in short supply.

Following from this I have two recommendations going forward. First, Canada should build a level of flexibility into its funding policies that allow for likely electoral hold-ups. Given Canada's long experience of working in Haiti, it should concede that electoral crises are not extraordinary but are part of the maturing process and expect delays that will occasionally lead to interruptions in the constitutional order. The democracy promotion literature has shown that democratic norms and institutions establish themselves slowly and the trajectory is not linear but rather fitful and irregular. Even if the electoral rules exist on paper, it can take many decades before they are faithfully adhered to. All good reasons for Canada to anticipate and prepare for electoral postponements.

A second recommendation would be that Canada remain remote and detached from these deeply political electoral disputes. I am certain this is Ottawa's intention. Nevertheless, in partnership with the United States, France, and the OAS, it has helped engineer resolutions to various electoral and political crises – some of them quite controversial.[13] It has allowed itself to be drawn into the country's political arena far too often, particularly during elections. While these mediations may have served to break an inconvenient political logjam, they may have hindered the development of Haiti's democratic norms and institutions. Moreover, the repeated political interventions of Canada and other outsiders have made Haitians deeply suspicious of foreign actors' motives, fuelling anti-foreign sentiment, which is understandable given a long history of "gross interference" in Haiti's domestic affairs – too long to detail here (Fatton 2014, 94).

Ottawa does not seem to accord this anti-foreigner sentiment sufficient consideration when deploying elections aid or crafting responses to Haiti's electoral crises. For instance, Canada (in partnership with the United States) has supported the capacity-building and training of a local election observer NGO, the Citizen Observatory for the Institutionalization of Democracy (OCID). Indeed, distinguished Canadian elections expert Neil Nevitte made numerous trips to Haiti to assist the group and support its observation work during the 2015 elections. Unfortunately, the aid OCID has received has affected its reputation. During the electoral crisis, stories in Haiti's media describing OCID's observation work almost always used the descriptor "supported by the US" or "US-funded," to its name. Indeed, graffiti following the October vote stated: "OCID + NDI = PHTK." The message was that OCID is funded by the US government, which supports the ruling party and its candidate, Jovenel Moïse. As a result, OCID's work and analysis, no matter how professional and competent, is not always viewed as politically neutral or objective – a serious problem given that it observes elections. In the future, Canada should consider distributing this capacity-building aid equally among all credible local observer groups, rather than funding just one. This will prevent the legitimacy, neutrality, and integrity of the sole recipient from being damaged.

Canadian Democracy Assistance: Going Forward

This case study of the 2015 electoral crisis highlights a general problem with Canada's approach to democracy assistance. Canada often places too high a value on fostering order and stability, allowing this first-order priority to shape its responses to political crises. This is wrong-headed and ultimately unworkable if democratization is indeed the goal. There are two reasons for this. The first relates to institution building.

The institutional modelling approach Canada employs, which is largely based on the notion that democracy can be learned with reference to examples, fails to fully appreciate the profoundly political nature of this undertaking. As these institutions take shape, they are not socially neutral. In other words, there are distributional consequences

(Grugel and Bishop 2014). Consequently, the stakes can be very high as social forces jostle to have their interests reflected in the new structures. Indeed, the contour and content of these political institutions is almost always the result of "a political settlement based on the prevailing power dynamics in society" (Grugel and Bishop 2014, 111). Forging that political settlement, as Haiti's various attempts to establish a permanent electoral council have shown, can take a very long time. Canada should recognize the central role of politics in the process of institution building, understand it involves real struggles that can be messy, even violent.

A second reason why too great a focus on stability is wrong-headed is because, as Grugel and Bishop have observed, for democracy to take hold "groups with interests embedded in the maintenance of the non-democratic status quo have to be either defeated or reformed" (Grugel and Bishop 2014, 4). In Haiti, a country with deeply entrenched authoritarian legacies, and where politics are weakly institutionalized, it seems unrealistic to expect policies that challenge privilege and redistribute political and economic power to slide into place smoothly, without resistance. Choosing to prioritize stability could have the effect of stalling or blocking these changes and the introduction of equal citizenship.

There is one final point. Democracy, when applied to Haiti, has long been imagined too narrowly. While Canada's understanding goes beyond the basic minimalist definition towards a more "rights-based" approach, it fails to recognize the issue of power. More specifically, insufficient attention is given to how socio-economic or other forms of structural inequalities prevent political participation.[14] Indeed, when Aristide was first elected in 1990, his efforts to redistribute power provoked an elite backlash, which produced a military coup. This serves to remind us that, in countries like Haiti, democratization is not simply the establishment of institutions and elections. At its heart it is a struggle to transform state structures and the relationship between the state and vulnerable and marginalized groups in society (Fatton 2014, 2007, Mintz 1995). This is a radical project. Canada can support it more effectively by recognizing it as such and adjusting its expectations during electoral and political crises accordingly.

NOTES

1 For an assessment of democracy promotion in the Americas region, see Thede 2002, 2005; Schmitz 2004; Daudelin 2007; and Burron 2011, for the Harper years, see Burron 2011 and Legler 2012.

2 For an excellent account of Canadian democracy promotion efforts during the 1990s, see Tom Keating, "Promoting Democracy in Haiti: Assessing the Practical and Ethical Implications," in Rosalind Irwin (ed.) *Ethics and Security in Canadian Foreign Policy* (Vancouver: UBC Press, 2001).

3 A comprehensive account of the events leading up to Aristide's ouster in 2004 are provided in: International Crisis Group, "A Chance for Haiti?" Latin American/Caribbean Report No. 10, (Port-au-Prince/Brussels: ICG, 18 November 2004). For an excellent account of the US role, see Walt Bogdanich and Jenny Nordberg, "Mixed U.S. Signals Helped Tilt Haiti toward Chaos," *The New York Times*, 29 January 2006.

4 Ottawa considers a state fragile when its government is unwilling or unable to carry out key functions, such as the enforcement of security, the protection and promotion of human rights and gender equality, the rule of law, and the basic provision of services (CIDA 2005, 6).

5 The Citizen Observatory for the Institutionalization of Democracy (OCID) is a Haitian election observation coalition created in 2015 by three civil society organizations: *Initiative de la Société Civile* (ISC), *Centre Oecumenique des Droits Humains* (CEDH), and *JURIMEDIA*. It carried out pre-election environment monitoring and observed the 9 August 2015 (first-round legislative) and 25 October 2015 (second-round legislative and local elections and the first round of presidential) votes. It deployed over 1,700 trained observers (US Department of State 2016) to monitor both rounds. Canada and the United States financially supported the establishment of OCID.

6 Author's translation. For the original French text, see Réseau National de Défense des Droits Humains, Conseil National d'Observation des Elections, and Conseil Haïtien des Acteurs non Etatiques 2015, 5.

7 The RNDDH was joined by three other local NGOs: *Solidarité Fanm Ayisyèn* (SOFA), *Conseil National d'Observation des Elections* (CNO), and *Conseil Haïtien des Acteurs Non Etatiques* (CONHANE).

8 US-based diaspora groups were also calling for a complete recount of all ballots by independent observers.

9 For a good summary of the report, see Elections Blog 2016 article on "ambiguous findings" (Johnston 2016).

10 The original 105-page Report in French can be found at: https://drive .google.com/file/d/0BwrRcOqQep6dOFlPaUgzLUxLWHM/view.

11 While the October vote also included elections for the Senate and the House of Deputies, the Commission did not recommend a rerun of those contests.

12 EU-Observers were funded mainly by the United States, Canada, and Brazil. In the end, the EU did not send observers to the redo, while the OAS did.

13 For accounts of how international actors, including Canada, have involved themselves in Haitian politics during the 2000, 2004, and 2008 electoral crises, see Engler and Fenton 2005; Dupuy 2007; Hallward 2007; Fatton 2011, 2015; and Seitenfus 2015.

14 This more limited understanding of democracy has led Canada to be less effective in its human rights and democratization work at times. For instance, Canada has vigorously supported women's political rights in Haiti. However, it has done little regarding labour rights (which would speak to resource redistribution), when the large majority export factory workers are women.

REFERENCES

Baranyi, Stephen. 2007. "Le Canada, Haiti et les Dilemmes de L'intervention dans les 'Etats Fragiles'." Ottawa: Paper presented at the Latin American Studies Association Congress in Montreal, 5–8 September 2007.
– 2008. "Canada and the Challenges of Democratic Development in Haiti, Notes for Presentation on Governance Panel." Focal/Carleton/DFAIT Conference on Canada and the Americas, 13–14 March 2008.
– 2010. "Canada and the Travail of Partnership in Haiti "In *Haiti's Governance Challenges and the International Community*, edited by Jorge Heine and Andrew Thompson, 205–28. Waterloo: Centre for International Governance Innovation (CIGI).
– 2012. "Contested Statehood and Statebuilding in Haiti." *Revista de Ciencia Politica* 32, no. 3: 723–38.
Brown, Stephen. 2018. "All about That Base? Branding and the Domestic Politics of Canadian Foreign Aid." *Canadian Foreign Policy Journal* 24, no. 2:145–64.
Burron, Neil. 2011. "Reconfiguring Canadian Democracy Promotion." *International Journal* (Spring): 391–417.
Burron, Neil, and Ray Silvius. 2013. "Low-Intensity Democracy and Political Crisis in Haiti: The North American Contribution." *Canadian Journal of Development Studies* 34, no. 4: 518–32.
Cameron, Maxwell, and Catherine Hecht. 2008. "Canada's Engagement with Democracies in the Americas." *Canadian Foreign Policy* 14, no. 3: 11–28.
Cammack, Paul. 2005. "The Governance of Global Capitalism: A New Materialist Perspective." In *The Global Governance Reader*, edited by Rorden Wilkinson, 156–73. London and New York: Routledge.
Carothers, Thomas. 1999. *Aiding Democracy Abroad. The Learning Curve.* Washington: Carnegie Endowment for International Peace.
Charles, Jacqueline. 2015a. "Concerns Widen over Haiti Presidential Elections Impasse." *Miami Herald* (30 November). Accessed 25 October 2021. https://

www.miamiherald.com/news/nation-world/world/americas/haiti
/article47129035.html.

– 2015b. "Haiti Opposition Rejects Martelly's Election Commission." *Miami Herald* (18 December). Accessed 25 October 2021. https://www.miami herald.com/news/nation-world/world/americas/article50215830.html.

– 2015c. "Haiti to Redo Legislative Elections in 25 Constituencies." *Miami Herald* (20 August). Accessed 25 October 2021. https://www.miamiherald .com/news/nation-world/world/americas/haiti/article31660835.html.

– 2016. "Haiti Panel Calls for Re-run of Presidential Elections." *Miami Herald* (30 May). Accessed 25 October 2021. https://www.miamiherald.com/news /nation-world/world/americas/haiti/article80825277.html#storylink=cpy.

CIDA. 2004. *Canadian Cooperation with Haiti: Reflecting on a Decade of "Difficult Partnership."* Ottawa: Canadian International Development Agency.

– 2005. *On the Road to Recovery: Breaking the Cycle of Poverty and Fragility, Guidelines for Effective Development in Fragile States.* CIDA.

– 2007. "Report on Plans and Priorities Estimates 2007–2008." Canadian International Development Agency. Accessed 25 October 2021. http:// www.tbs-sct.gc.ca/dpr-rmr/2006-2007/inst/ida/ida01-eng.asp.

Daudelin, Jean. 2007. "Canada and the Americas: A Time for Modesty." *Behind the Headlines* 64, no. 3: 1–31. Toronto and Waterloo: Canadian Institute for International Affairs and the Centre for International Governance Innovation.

Diamond, Larry. 1993. "Introduction: Political Culture and Democracy." In *Political Culture and Democracy in Developing Countries*, edited by Larry Diamond, 1–33. Boulder, CO: Lynn Rienner.

Dupuy, Alex. 1994. "Free Trade and Underdevelopment in Haiti: The World Bank/USAID Agenda for Social Change in the Post-Duvalier Era." In *The Caribbean in the Global Economy*, edited by Hilbourne Watson, 91–107. Boulder and Kingston: Lynne Rienner Publishers & Ian Randall Publishers.

– 2007. *The Prophet and Power.* Lanham, MD: Rowman & Littlefield.

Dyer, Evan. 2016. "Canada Showing Haiti Some Tough Love." CBC News (22 November). Accessed 25 October 2021. https://www.cbc.ca/news /politics/haiti-elections-canada-1.3769376.

Engler, Yves, and Anthony Fenton. 2005. *Canada in Haiti: Waging War on the Poor Majority.* Halifax: Fernwood.

Euractiv.com and Agence France Press (AFP). 2016. "EU Bitter after Haiti Cancels Election Results." Accessed 25 October 2021. https://www .euractiv.com/section/global-europe/news/eu-bitter-after-haiti-cancels -election-results/.

Fatton, Robert Jr. 2007. *The Roots of Haitian Despotism.* Boulder, CO: Lynne Rienner.

– 2011. "Haiti's Unending Crisis of Governance: Food, the Constitution and the Struggle for Power." In *Fixing Haiti*, edited by Jorge Heine and Andrew Thompson, 41–65. New York: United Nations University Press.

– 2014. *Haiti's Neoliberal Trap: Living in the Outer-Periphery*. Boulder, CO: Lynne Rienner.

– 2015. "Killing Haitian Democracy." (22 July). Accessed 25 October 2021. https://www.jacobinmag.com/2015/07/monroe-doctrine-1915-occupation -duvalier/.

Fenton, Anthony. 2004. "The Canadian Connection." ZNet (11 March). Accessed 25 October 2021. https://zcomm.org/znetarticle/the-canadian -connection-by-anthony-fenton/.

Global Affairs Canada. 2017. "Statistical Report on International Assistance 2016–2017." Government of Canada. Accessed 25 October 2021. https:// www.international.gc.ca/gac-amc/assets/pdfs/publications/sria-rsai-2016 -17-eng.pdf.

Grugel, Jean, and Matthew Louis Bishop. 2014. *Democratization: A Critical Introduction*. Hampshire and New York: Palgrave Macmillan.

Guyler Delva, Joseph. 2016. "Haiti Opposition Rejects President's Plan for Interim Government." Reuters (2 February). Accessed 25 October 2021. https://www.reuters.com/article/us-haiti-election-idUSKCN0VB0PY.

Haiti Advocacy Working Group. 2016. "Panel Discussion: Supporting Free & Fair Elections: Perspectives from Haiti, the U.S. and the International Community. March 4, 2016." (2 March). Accessed 25 October 2021. https:// haitiadvocacy.org/supporting-free-fair-elections-haiti/.

Hallward, Peter. 2007. *Damming the Flood. Haiti, Aristide, and the Politics of Containment*. London: Verso.

Ives, Kim. 2016. "Finding Only 9% of Votes Valid: Haiti Verification Commission Says Presidential Election First-Round Should be Scrapped." *Haiti Liberte* (7 June). Accessed 25 October 2021. http://www.haiti-liberte .com/archives/volume9-47/Haiti%20Verification%20Commission.asp.

Johnston, Jake. 2016. "Evaluation Commission's Ambiguous Report May Only Deepen Haiti's Electoral Crisis." (5 January). Accessed 8 December 2021. https://cepr.net/evaluation-commission-s-ambiguous-report-may-only -deepen-haiti-s-electoral-crisis/.

Keating, Tom. 2001. "Promoting Democracy in Haiti: Assessing the Practical and Ethical Implications." In *Ethics and Security in Canadian Foreign Policy*, edited by Rosalind Irwin, 208–26. Vancouver: UBC Press.

Legler, Thomas. 2012. "Wishful Thinking: Democracy Promotion in the Americas under Harper." *International Journal* (Summer): 583–601.

Mintz, Sidney. 1995. "Can Haiti Change?" *Foreign Affairs* 74, no. 1: 73–86.

Muggah, Robert. 2007. "The Perils of Changing Donor Priorities: The Case of Haiti." In *Exporting Good Governance: Temptations and Challenges in Canada's*

Foreign Aid Program, edited by Jennifer Welsh and Ngaire Woods, 169–202. Waterloo: The Centre for International Governance Innovation/Wilfrid Laurier University Press.

Noriega, Roger F. 2016. "Lawless Leaders Threaten Haiti's Future." Foreign Policy(16 June). Accessed 25 October 2021.https://foreignpolicy.com/2016/06/06/lawless-leaders-threaten-haitis-future/.

Norris, Pippa. 2013. "The New Research Agenda Studying Electoral Integrity." *Electoral Studies* 32: 563–75.

OAS. 2017. "Report to the Permanent Council. Electoral Observation Mission." Presented by Ambassador Juan Raul Ferreira for the presidential, legislative, municipal, and local elections of 20 November 2016 and 29 January 2017. Washington, DC: Organization of American States.

Observatoire Citoyen pour l'Institutionnalisation de la Démocratie (OCID). 2015. Diagnostic de l'engagement citoyen dans la perspective des elections de 2015 en Haïti. Rapport d'une enquête nationale de l'Observatoire Citoyen pour l'Institutionnalisation de la Démocratie (OCID). Sous la direction de Neil Nevitte. Port-au-Prince: OCID.

Patriquin, Martin. 2006. "Accusing Canada of Corruption *"Maclean's* (6 March), 19.

Permanent Mission of Canada at the OAS. 2016. Note from the Permanent Mission of Canada Requesting the Inclusion of the Topic "The Electoral Process In Haiti," in the Draft Order of Business for the Next Meeting of the Permanent Council. Washington, DC.

Réseau National de Défense des Droits Humains, Conseil National d'Observation des Elections, and Conseil Haïtien des Acteurs non Etatiques. 2015. "Communiqué de Presse: Scrutin du 9 août 2015: un accroc aux normes démocratiques." Accessed 8 December 2021. https://www.slideshare.net/RadioTelevisionCaraibes/scrutin-du-25-octobre-rapport-de-la-solidarite-fanm-ayisyn-sofa-le-conseil-national-dobservation-des-elections-cno-le-conseil-hatien-des-acteurs-non-etatiques-conhane-et-le-rseau-national-de-dfense-des-droits-humains-rnddh.

Reuters. 2016. "Haiti to Fund $55-Million New Try at Elections after Results Scrapped." (17 August 17) Accessed 25 October 2021. https://www.reuters.com/article/us-haiti-election/haiti-to-fund-55-million-new-try-at-elections-after-results-scrapped-idUSKCN10T0CG.

Robinson, William. 1996. *Promoting Polyarchy. Globalization, US Intervention and Hegemony.* Cambridge: Cambridge University Press.

Sanders, Richard. 2007. "A Very Canadian Coup D'état in Haiti: The Top 10 Ways That Canada's Government Helped the 2004 Coup and its Reign of Terror." *Press for Conversion* (60).

Sanders, Ronald. 2016. "In Haiti, There Is No Quick Fix after Postponed Presidential Elections." (3 August). Accessed 25 October 2021. https://www.antillean.org/haiti-presidential-elections-second-round13810-2/.

Schmitz, Gerald. 2004. "The Role of International Democracy Promotion in Canada's Foreign Policy." *IRPP Policy Matters* 5, no. 10.

Seitenfus, Ricardo. 2015. *L'échec de l'aide internationale à Haïti: Dilemmes et égarements* Port-au-Prince: Éditions de l'Université d'État d'Haïti.

Shamsie, Yasmine. 2006. "It's Not Just Afghanistan or Darfur: Canada's Peacebuilding Efforts in Haiti." In *Canada Among Nations 2006: Minorities and Priorities*, edited by Andrew F. Cooper and Dane Rowlands, 209–31. Montreal and Kingston: McGill-Queen's University Press.

– 2008. "Canada's Approach to Democratization in Haiti: Some Reflections for the Coming Years." *Canadian Foreign Policy* 14, no. 3: 87–93.

– 2009. "Export Processing Zones: The Purported Glimmer in Haiti's Development Murk." *Review of International Political Economy* 16, no. 4: 649–72.

Simpson, E., and Brian Tomlinson. 2006. "Canada: Is Anyone Listening?" In *Reality of Aid 2006: Focus on Conflict, Security and Development*, 256–61. New York: Zed Books.

Thede, Nancy. 2002. *Democratic Development 1990–2000: An Overview*. Montreal: International Centre for Human Rights and Democratic Development.

– 2005. "Human Rights and Democracy: Issues for Canadian Policy in Democracy Promotion." *IRPP Policy Matters Series* 6, no. 3.

UN News Service. 2016. "UN Chief: Haiti Can 'Ill Afford' Prolonged Period of Transitional Governance." (2 June) Accessed 25 October 2021. https://news.un.org/en/story/2016/06/530992-un-chief-haiti-can-ill-afford-prolonged-period-transitional-governance.

Union Europeenne Mission d'Observation Electorale Haïti 2015 Election Générales. 2016. *Analyse MOE UE du rapport de la Commission Indépendante d'Evaluation et Vérification Electorale (CIEVE)*.

US Department of State. 2016. "Attachment H, Report." accessed 3 August 2018. https://www.state.gov/p/wha/ci/ha/hsc/2016report/266874.htm.

6 Latin American Migration to Canada: New and Complex Patterns of Mobility

CHRISTINA GABRIEL AND LAURA MACDONALD

In this chapter we will explore the shifting nature of migration from Latin America to Canada. There is relatively little literature on this phenomenon (Hernández-Ramírez 2019). This gap may reflect the relatively small size of the Latin American community in Canada compared to other migrant groups and also some of the features of that diaspora: It is relatively dispersed, non-cohesive, and less prosperous than some other immigrant communities. Large-scale migration to Canada is also relatively recent, not having begun until the 1970s, while the United States has often represented the destination of choice for Latin American migrants (Simmons 1993). These features also mean that Latin American Canadians have lacked political influence in Canada, compared to some other immigrant groups. Nonetheless, Latin Americans represent an important migration flow to Canada, especially through their participation in the Temporary Foreign Worker Program (TFWP).[1] Canada is the third most important destination for Latin American and Caribbean emigration, after the United States and Spain (SICREMI 2011: vii). As this volume seeks to look beyond traditional state-centric accounts of the Canada–Latin America relationship, it is vital to include an analysis of the nature of migration between the two areas, how it has changed over time, and how we understand changes in this relationship.

The existing academic literature on Latin American migration to Canada tends to discuss the phenomenon in terms of successive "waves" of migrant groups who have come to Canada, often fleeing political and military persecution or economic marginalization in their home communities (Mata 1985; Simmons 1993; Hernández-Ramírez 2019). Thus, movements of peoples are often viewed through the familiar frame of "push" factors in countries of origin and "pull" factors from destination sites.

We argue in this chapter that changes in the global political environment coupled with changes in Canadian immigration policy and new migration patterns mean that this portrayal of the conditions that shape migration may no longer accurately describe the conditions that shape Latin Americans' migration to Canada (see Veronis 2007, 457). We see new patterns and conditions that are reconfiguring the migration landscape. These include: a greater prevalence of temporary migration; more sustained and continued cultural, economic, and social exchanges between "home" and "host" locations, and new, sometimes contradictory, policies in both Canada and the United States that have changed the nature of migrants' journeys. These new policies have led to new pathways and categories of migration and include an increased emphasis on temporary migration of skilled migrants under new free trade agreements. At the same time, new control and bordering practices have also intensified in this period that have made some migrants' journeys to Canada more difficult than in the past. The 2002 Canada–United States Safe Third Country agreement and the imposition of a visa requirement on Mexicans under the Harper government (see Gutiérrez Haces's chapter in this volume) have limited the numbers of Latin Americans fleeing official persecution or gang violence who can claim refuge in Canada. The anti-migrant policies of the Trump administration may have also created incentives for documented and undocumented Latin Americans to move to Canada from the United States; it is unclear how the migration policies of the Biden administration will affect these dynamics.[2] All of these factors have complicated the movements of Latin Americans to Canada but also highlight the importance of migration in shaping Canada–Latin America relations.

We begin the chapter by briefly considering some of the frames and conceptual metaphors that have been mobilized to describe Latin American migration to Canada before moving on to an overview of the history of these movements. We discuss in this section the traditional depiction of Latin American migration to Canada as taking the form of "waves" of migrants, in which the majority of migrants were individuals fleeing violence and political oppression in their countries of origin. In the second half of the twentieth century, Canada removed its earlier explicit race-based exclusionary policies and also began to open its doors to victims of repression by regimes supported by the United States. In the next section we discuss the shifting character of migration in the current century and emphasize the increased complexity of migrant motives and identities but also the ways in which changes in Canadian policies have complicated the ways in which Latin Americans can gain access to Canada. In this period, the Canadian state has adopted new policies that attempt to increase the economic benefits it receives from migration but also has engaged in different forms of "bordering" that restrict migrant entry on the basis of perceived threat and limit the potential of asylum claimants to make their claims in Canada. Here we focus first on the rise in temporary forms of mobility, both high- and low-skill, which have increased Latin Americans' access to the Canadian labour market, but only for limited periods. Moreover, these temporary migrant programmes are often associated with high levels of insecurity and vulnerability. While these programmes have increased Latin Americans' entry into Canada, other policies have restricted their entry. In this context we discuss the Safe Third Country Agreement, signed between the United States and Canada in 2002, which limits the ability of asylum claimants to file a refugee claim in Canada if they previously passed through the United States, considered a "safe" site for them to file a claim. We also describe the politics of the Canadian imposition of a visa requirement on Mexicans under the Harper government, a requirement that was subsequently lifted by the Trudeau government. We conclude by arguing that an understanding of the shifting character of Latin American migration to Canada must move beyond conventional analytical frameworks based on the image of "waves" of migrants to more nuanced approaches that take into account both changes in the Latin American region and political changes in Canada.

Frames and Conceptual Metaphors: The Analytics of Latin American Migration to Canada

Scholars seeking to interrogate migration patterns and processes have adopted various markers to understand these dynamics. In the wake of ongoing changes in migration governance, the expansion of temporary/circular migration, new border practices, and the rise of populist anti-immigration rhetoric, the familiar frames and conceptual metaphors that guided earlier accounts of Latin American migration to Canada need to be revisited. This section briefly highlights common markers that have been used in earlier assessments of flows from Latin America to Canada.

The metaphor of successive waves has been used to periodize and map Latin American migration to Canada (Mata 1985; Simmons 1993; Goldring 2006; Landolt et. al 2011). This typology developed by Mata (1985) focuses on the period of entry of Latin Americans as permanent immigrants and is attuned both to the sociopolitical environments in sending countries as well as to the immigration policy infrastructure in Canada. Thus, Latin American migration is presented as a series of four waves. For instance, the first wave occurs with the arrival of a small number of Latin Americans with European origins in the 1960s, while the fourth wave comprises refugees fleeing from Central America. Scholars have extended this metaphor. A fifth wave has been added to capture Latin American migration in the mid-1990s, "characterized by a diversification of countries of origin, contexts of departure, and modes of migration and entry," including tourists, asylum seekers, temporary workers, and middle-class entrepreneurs and professionals (Landolt et. al. 2011, 1243). The focus in the discussion of "ideal type" waves is, with the exception of the fifth wave, on permanent migrants to Canada and the factors that motivated them to migrate from their countries of origin. To what extent does this approach sideline migration dynamics that do not fit this framework? Can it account for more complex patterns? Landolt et al. noted that "Each wave of Latin American migration has been fairly short-lived and linked to a specific political event or economic situation in the country of origin" (2011, 1243). It is not clear that this observation holds up today, and it might be difficult to offer a straightforward wave categorization given the increasing diversity, heterogeneity, and complexity associated with current Latin American flows to Canada that we discuss in the later portion of the chapter.

To the extent that the wave metaphor references factors that influence migrants to leave their countries of origin, it resonates with the more familiar immigration "push-pull" framework that addresses asymmetries between countries of origin and destination. In classic formulations of this framework, people would be prompted to move on the basis of external economic factors – from poorer to more prosperous countries or from low wage to high wage regions. But this formulation has expanded to include a broader range of factors (O'Reilly 2016, 25–6). As Castles et. al. detail: "Push-pull models identify economic, environmental, demographic factors which are assumed to push people out of places of origin and pull them into destination places. 'Push factors' usually include population growth and population density, lack of economic opportunities and political repression, while 'pull factors' usually include demand for labour, availability of land, economic opportunities and political freedoms" (2014, 28). They highlight that the model appears to be attractive insofar as it draws together the many factors that may be implicated in processes of migration. However, they point out that the model is largely descriptive because it falls short in its ability to explain the role and significance of these factors or the ways in which they interact (2014, 28–9). Others suggest that this type of framework was static in character insofar as it "failed to account for changing motivations, altered circumstances or modified decisions en route" (Van Hear et al. 2018, 929, citing de Haas 2011). Further, push-and-pull frames tend to overlook how individuals' decisions to migrate are constrained by broader structural conditions (Castles et al. 2014, 31).

In his account of Latin American migration, Simmons references the conceptual metaphor of "waves" and a categorization that gestures towards push–pull factors. But he locates these factors within a broader structural context by drawing our attention to what he terms "a hemispheric migration system" (1993, 284). Importantly, he notes that the future migration patterns will be shaped by "the North American Free Trade Agreement; the state of peace and democracy in Central America; the potential for new insurgency and violence; [and] changes in Canadian immigration and refugee policy." He concludes by presciently observing that all these elements will be shaped by broader global processes (1993, 305). And indeed it has been observed that the effects of globalization on migration have to be seen through changes in the organization of production, the structure of labour markets, and social disparities. Relatedly, the role of the state in migration governance, though changing, has not disappeared. Increasingly, in the North, we see the construction of differentiated migration infrastructures designed to facilitate and attract the movement of high-skilled individuals while often discouraging the movement of low-skilled workers and asylum seekers (Castles et al. 2014, 34-5). The former often have an easier route to permanent status and citizenship, while the latter are often excluded altogether or occupy a range of precarious statuses.

Simmons also highlights the importance of migrant networks: "once a migration stream has been established by a few trailblazing pioneers, other relatives, friends and members of the same community will tend to follow" (1993, 304). Consequently, some scholars have addressed the role of Latin American migrants in Canada by considering processes of integration and migrant links to countries of origin through a transnational framework (Goldring 2006; Veronis 2010). As Goldring and Landolt put it, "transnational social fields and related analytical categories—networks, practices, and identities that span nation-state borders—are the conceptual starting point for studying social relations, formations and processes … that are not necessarily bound by place or national containers" (2014, 106). The emphasis on transnational entanglements of Latin American groups in Canada differs in focus from the literature that attempts to problematize migration flows.

The scholarly interventions discussed here (Mata 1985; Simmons 1993) have addressed Latin American migration to Canada in terms of the broader issue of what initiated a migration process. To some extent they are underpinned by the assumption that this migration would be permanent and not temporary or circular. There is also an implicit assumption that people making claims as refugees can be clearly differentiated from economic migrants. The increasing diversity of Latin American migration to Canada and broader transformations seems to suggest that "wave" may no longer be an apt metaphor. We discuss the historical evolution of migration from Latin America to Canada in the next section.

Historical Background

Prior to 1970, Latin American migration to Canada was extremely limited, reflecting the lack of strong social and cultural ties between the two areas, even if some economic linkages did exist (Ogelsby 1976). Studying Latin Americans' presence in Canada is complicated by the heterogenous and contested nature of the term "Latin American." The geographic designation "Latin America" is a colonial construct, first applied by the French in their attempt to claim influence in the New World based on shared use of Latin languages. It is now generally used to refer to Spanish- and Portuguese-speaking countries in North, Central, and South America and the Caribbean. This definition, however, ignores and marginalizes the presence of large numbers of Indigenous language-speakers who first occupied this territory and lost control over much of it to colonial predation but continue to maintain an important presence today. It also leaves unclear and disputed some sites such as Haiti, which is often included in the category "Latin America" despite the dominance of the French and Créyol languages in that country, as well as other French-speaking Caribbean colonies which are not usually included. The term "American" is also contested. As decolonial historian Walter Mignolo argues, the idea of "'America' as we know it was an *invention* forged in the process of European colonial history and the consolidation and expansion of the Western world view and institutions" (2005, 2, his emphasis). Many migrants from the region known as Latin America would not think of themselves as "Latin American," but rather Colombian, Guatemalan, Chilean, Argentine, etc., or may even identify with a subnational region or city or Indigenous nationality (Ginieniewicz and McKenzie 2014, 263–4).

Once in Canada, these individuals have not normally formed a cohesive identity or culture, and fragmentation occurs even within the same national origin group because of class, ethnic/racial, cultural, age, gender, and political differences, as well as their geographic dispersion within Canada (Armony 2014, 14; Goldring and Landolt 2009; 2010), mostly located in the three largest cities of Toronto, Montreal, and Vancouver. Nevertheless, as argued by Luisa Veronis (2007, 461),

Latin Americans do represent a cohesive, homogenous category in the eyes of the Canadian state, and, partly in response, Latin Americans in Canada may adopt a form of "strategic essentialism" to promote shared objectives and interests. Several groups representing Latin American populations, such as the Canadian Hispanic Congress (CHC), have encouraged Statistics Canada to include factors such as "country of birth" and "ethnic identity" to better identify community members (see also discussion in Armony 2014, 18–20).

Latin American migration to Canada did not reach statistically significant numbers until the mid-twentieth century, both because of Latin Americans' lack of knowledge of and interest in Canada and the greater proximity of the large US labour market, but also because of Canadian restrictions on immigration of non-European settlers. In the 1960s Canada removed its exclusionary policies based on "preferred country" source of migrants that effectively prohibited flows from Latin America. Canada instead adopted a "points system" that overall favoured skilled economic migrants, in contrast with the United States' greater emphasis on family unification, which eventually resulted (along with geographic proximity) in large numbers of Mexican migrants. While Mexicans were by far the largest category of Latin American immigrants in the United States, Canada's Latin American population became much more diverse in terms of country of origin, with three main immigrant groups (Mexican, Colombian, and Salvadorian) (Armony 2014, 18).

As mentioned above, in an influential 1985 article, Fernando Mata describes four "waves" of Latin American migrants that followed the Canadian government's policy change. In any given period, the majority of migrants came from only a few Latin American countries to Canada. The first, or leading, wave was small and consisted largely of educated professionals, often born in Europe, and arrived from Latin America in a process of step migration in the early 1960s. A second wave came later in the 1960s from Ecuador, Peru, and Colombia in response to increased demand for both skilled and unskilled labour in Canada's industrial heartland. Numbers in this period were still extremely small (Mata 1985; Goldring and Landolt 2009, 1232). The third wave began during the 1970s and included people fleeing new military dictatorships in the South American region, while the fourth wave, during the 1980s, was made up of individuals fleeing military repression and low-intensity warfare, sponsored by the United States, in Central America. Berdichewsky (2007) identifies a fifth wave of Mexicans entering Canada during the mid-1990s, entering under the economic and family migration classes. Hernández-Ramírez (2019, 80) argues there were subsequently a sixth wave of Colombians and Mexicans who immigrated between 2000 and 2009 and a seventh wave that developed between 2010 and 2016, in which Mexicans ranked first and Colombians second.

Simmons argues that the timing and size of the flows, which became more significant after 1970, responded to several factors:

a) social, economic and political conditions in the sending countries;
b) social, economic, and political conditions in destination countries including Canada;
c) Canadian immigration and refugee policies, which changed first to permit immigrants, and later changed to admit refugees;
d) migrant social and kin networks (Simmons 1993, 285).

He also argues that we can understand migration in the Americas as functioning as a regional system, like a wheel, with the United States at the hub and other countries in the region as spikes. In general, the United States is the preferred destination for migrants from all of the other countries, followed by migration between geographically neighbouring countries. Canada usually attracts much smaller numbers of migrants, as the "next best" North American destination (1993, 289). However, in certain periods, especially during the Cold War, Canada became a more preferred choice for asylum seekers because of its greater openness to immigrants and refugees from U.S. allies. This argument implies that economic motivations were not the primary cause of mobility during the early waves and that the migrants were often victims of state-sponsored terror. Even those individuals who did leave for primarily economic motivations often chose to leave their countries of origin because of the context of political instability and violence.

This dynamic became evident when larger-scale migration began in the early 1970s after a series of military dictatorships came to power in several South American countries. Perhaps the most significant case for Canada was that of the Chilean military coup of 1973, which overthrew the democratically elected government of socialist president Salvador Allende and brought to power the brutal right-wing dictatorship led by General Augusto Pinochet, with the backing of the United States. The Canadian government decided to recognize the new government, despite fierce opposition from the left-wing New Democratic Party and Canadian civil society actors, including efforts by trade unions, church groups, and other humanitarian lobbies. The Canadian government had not been very sympathetic to the Allende government and already had a practice of recognizing military governments (Stevenson 2000, 127). After NGOs lost this battle, they turned to pressuring the Canadian government to admit Chilean supporters of Allende who were endangered by the new regime, even though Canada's refugee policy was not formalized until the passage of the 1976 Immigration Act (Simmons 1993; García 2006, 122).

The Canadian embassy was under pressure from the United States not to admit significant numbers of asylum seekers and was pushing Ottawa to slow down the processing of refugees, apparently believing that the number of Chileans who chose Canada as a destination would not be very high. However, under pressure from Parliament and civil society, the government chose to relax its approach to processing applications from refugees, and began a "Special Movement Chile," initially authorized to bring between 300 and 1,000 Chileans to Canada, a number that was later increased, again as a result of NGO pressure, to 7,000 (Stevenson 2000, 127–30). Following this example, through the 1970s, other refugees were accepted from Argentina and Uruguay in response to military regimes' human rights abuses in those countries. In 1975, in the peak of this phase of mobility, around 10,000 migrants arrived from Latin America (Simmons 1993, 47). Canadian migration policies thus changed in this period to create new paths for refugees and marked the beginning of a gradual decline of Canadian deference to the United States in the Western Hemisphere. This episode also marked one of the most important examples of civil society pressure, resulting in a significant shift in Canadian foreign policy (Macdonald 2018). Once in Canada, these Latin American migrants joined with Canadian civil society actors to pressure the Canadian government to distance itself from US foreign policy and condemn the military dictatorships in the region (see the chapter by John Foster in this volume).

The pattern established in the 1970s with South Americans was repeated in the 1980s after the upsurge of left-wing revolutionary movements in Central America, and the United States' response in the form of support for right-wing dictatorships and counter-revolutionary forces. The resulting decade of civil wars led to large movements of people, fleeing from "dictatorship, state terror, civil war and violence" (Simmons 1993, 282). Again, Canada became a preferred destination for many Latin Americans, overcoming the historic hub-and-spoke model in which Latin Americans had historically favoured the United States as a destination. In the context of President Ronald Reagan's revival of Cold War policies, the United States refused to recognize most refugee claimants coming from its ally regimes in El Salvador, Honduras, and Nicaragua, while Canada took a more open approach. According to María Cristina García, Canada's refugee policy was the principal way in which the Canadian government differentiated itself from the policies of the United States in Central America (2006, 119). To accommodate refugee needs, the government gradually increased its refugee quota for Latin America from 1,000 in 1981 to 2,500 in 1984. In 1986, the quota for Latin America was set at 3,200, but Canada accepted 3,654 from Central America alone. Overall, between 1982 and 1987, Canada admitted 15,877 Central American refugees, the large majority (11,251) from El Salvador, and Salvadorans replaced Chileans as the largest Latin American community group in Canada (García 2006, 128–30).

In total, some 26,000 Salvadoreans arrived in Canada between 1973 and 1990 (Simmons 1993, 294). Many of these refugees had spent time in Mexico and/or the United States before coming to Canada. Like Mexico, Canada's refugee policy permitted the government to distinguish itself from US policy in the region (García 2006, 130). In 1986, however, the US Congress passed the Immigration Reform and Control Act, in part to reduce the number of undocumented migrants in its territory. This measure created a "rush" of Central American migrants leaving the United States to claim asylum in Canada. From December 1986 to February 1987, 10,000 refugees, mostly Central Americans, arrived in Canada, and churches on both sides of the border provided shelter for large numbers of migrants. This upsurge in refugee claimants resulted in the Canadian government imposing new restrictions to limit applications for asylum, such as lifting a moratorium on deportations and revocation of a migrant's automatic approval to seek work and receive social services while awaiting ministerial approval of his or her petition. These measures resulted in a rapid reduction in the number of claimants (García 2006, 131–3). By the end of the decade, the Central American peace process temporarily stabilized the situation in the region and reduced the flow of refugees.

The end of the Cold War, the decline of US hegemony in Latin America, and the emergence of neoliberalism led to new patterns of migration towards Canada. Since then, there has been a diversification of Latin American migrants' modes of entry, with a shift away from refugee movements and towards economic categories. This shift makes the earlier metaphor of "waves" less relevant. While most of the Central Americans who arrived had been of humble origins, the 1990s saw a new stream of higher-skilled professionals arriving to Canada from countries such as Mexico, Colombia, and Argentina (Goldring 2006). Other migrants arrived as part of the family class category and as refugee claimants, with significant numbers of Mexicans, Colombians, and Venezuelans still making claims in the 1990s and 2000s. This marked the launch of a new phase of migrant mobility between Latin America and Canada, marked by new motives on the part of both migrants themselves and the receiving state, described in the next section.

New Trends in the 2000s: From "Waves" to Multiple Forms of Mobility

As argued above, the multiple and contradictory changes that have occurred in the regional and global systems since the mid 1990s have resulted in new forms of migration that must be interpreted in different ways. Migration scholars have moved away from the "push/pull" model of migration and from more essentialized approaches to migrants that emphasize a movement from cultural distinctiveness to assimilation. Contemporary theorists draw instead upon concepts that emphasize the fluidity and multiplicity of individuals' identities, such as "hybridity" and "transnationalism" (Veronis 2007, 457).

In this chapter, we adopt the definition of migrant transnationalism provided by Goldring, Henders, and Vandergeest, who say that it represents "the process whereby social actors who have migrated maintain active ties with their homelands, or across national borders through participation in religious, social, cultural, economic and political networks and processes." They emphasize that this phenomenon is not new but that the contemporary manifestation of this process may differ from earlier ones in its scope, intensity, and frequency, as well as in the way it is studied by academics (2003, n.p.). Riaño and Goldring (2014, 4) emphasize several dimensions of this process of relevance to our analysis. Migrant transnationalism, they argue, is multidimensional, "including socio-cultural practices and identities, economic and political exchanges and networks and flows of information and ideas." It is multi-scalar, involving macro political-economic processes as well as supranational agencies, nation states, NGOs, and social networks, families, and individuals. Transnationalism coexists with processes of incorporation, with the most established migrants often most engaged in attempting to influence homeland affairs. Manifestations of transnationalism vary across and within groups, and they are shaped by migrants' different experiences of departure and reception in their new country. A series of policy changes adopted by the Canadian government in the early years of the twenty-first century have had significant implications for how many Latin Americans have arrived, for how long they can stay, and what their conditions are upon arrival. On the one hand, these policies have had the effect of diversifying the Latin

American migrant population in Canada, but on the other hand they have also limited the possibilities for many groups to attain permanent status in the country.

Expansion of Pathways for Temporary Mobility

One major trend in the character of Latin American migration to Canada is the increased diversity in Latin American migrants who are moving to Canada and the growth in the number of pathways they are able to pursue. Overall, there is a decline in the number of Latin Americans who travel to Canada as refugees (partly in response to the bordering practices described below); an increase in the number of migrants coming under the economic category; and, most significantly, an increase in the number of temporary migrants. There has also been a decrease in the percentage of Latin Americans who come to Canada as permanent residents, from 11 per cent of the total number of new permanent residents in 2007 to 7.3 per cent in 2016, reflecting the decline in overall levels of migration out of Latin America in the early twenty-first century (Government of Canada 2016a). Temporary forms of mobility have occurred both within long-established migratory channels, such as the Mexican workers admitted under the Seasonal Agricultural Workers Program (SAWP), and new forms of temporary mobility, such as the Temporary Foreign Workers Program (TFWP) and the International Mobility Program (IMP). We discuss each of these categories below. This shift reflects broader changes in the organization of capitalism in Canada, with the "proliferation of flexible production systems and transformations in the organization of work" (Goldring and Landolt 2011, 326). More flexible production systems encourage both increased precarity in the Canadian labour market and state and capital's search for more new vulnerable workers, involving not just low-skill but also high-skill workers.

Temporary migration is a complicated phenomenon and include both high-skill and low-skill workers. The increase in the number of temporary migrants has particularly affected Latin Americans. Latin Americans figure prominently in the low-skill temporary worker category, as the number one source destination for the TFWP, with 25,918 migrants coming under this category in 2016 (up substantially from 17,037 in 2007). Guatemala is ranked fourth, primarily because of that country's workers' participation in the low-skill pilot programmes, with 6,509 work permits issued in 2016. In contrast, Latin Americans figure much less prominently in the high-skill International Mobility Program (IMP): Among all source countries for the IMP, Brazil is ranked number twelve and is the highest Latin American source country, with 4,363 workers in 2016 (up from 849 in 2007), followed by Mexico, which is ranked thirteenth, with 2,668 workers (up from 1,591 in 2007; Government of Canada 2016b).

The increase in temporary migrant pathways reflects a broader shift in Canadian immigration policy. In 2014, under the Harper government, the admission of temporary workers was restructured to fall under two categories: the Temporary Foreign Worker Program (TFWP) and the International Mobility Program (IMP). The TFWP largely targets low-skill workers (as defined under the National Occupational Categorization [NOC] system). It include both the Seasonal Agricultural Worker Program (SAWP) and the Live-In Caregiver Program (LCP). It also include the so-called Low-Skill Pilot (LSP) programme, which was introduced in 2002, which permits the entry of low-skill individuals for both agricultural work and other lower-skill occupations, such as those in the fast-food industry. The IMP is designed to "advance Canada's broad economic and cultural national interest" (IRCC 2016, 13) and include three subcategories: significant benefit (aimed at entrepreneurs, the self-employed, and intra-company transferees); agreements (referring to nationals of countries with which Canada has signed trade agreements and who are qualified to enter temporarily under that agreement), and other work permits (including postdoctoral fellows, students, foreign medical residents, and former international students in Canada who qualify for postgraduate work permits (Prokofenko and Hou 2018a, 9–10). In 2015, there were approximately 310,000 temporary migrants resident in Canada, more than four times the number in 2000 (Prokofenko and Hou 2018b).

The conditions under which these various categories of temporary migrants enter and remain in Canada vary significantly, with higher-skilled individuals generally having greater opportunities to access Canadian citizenship than the lower-skilled workers entering under the TFWP. The workers coming into Canada under the LCP (which include few Latin Americans) are able to apply for permanent residence after several years of work in Canada (under sometimes exploitative conditions); however, the number of workers admitted under the LCP has declined by almost half since 2014 after the minister introduced new rules.[3] Workers admitted under the TFWP are only able to access permanent status under the Provincial Nominees Program, and SAWP workers are not even able to apply for the PNP. As a result, the numbers of migrants gaining permanent status is extremely lopsided: while in 2016, 34,661 IMP workers and 4,985 high-skilled TFWP workers became permanent residents, only 369 agricultural workers did so (Government of Canada 2016b).

The conditions for employment are also skewed in favour of higher-skill entrants. To hire a worker under the TFWP (including the SAWP), employers must first obtain a Labour Market Impact Assessment (LMIA) from Employment and Skills Development/Service Canada. The programme is thus designed, according to Immigration, Refugees and Citizenship Canada (IRCC) to "help fill genuine labour needs as a last and limited resort when qualified Canadians or permanent residents are not available." In contrast, the IMP is designed "to provide competitive advantages to Canada and reciprocal benefits to Canadians" (Government of Canada 2016b). Migrants arriving in low-skilled categories are tied to a specific employer, while those entering under the IMP and who are LMIA-exempt, receive "open" work permits that allow workers to move between employers once in Canada. Critics of the TFWP often point to this lack of mobility as a key element in fostering vulnerability among migrants, since employers can arrange for workers with whom they are unhappy to be repatriated to their country of origin; high-skill migrants arriving under the IMP do not experience this same form of vulnerability. Recipients of open work permits are also able to bring their spouses and family members with them, while other temporary workers are not (Fischer 2017). In general, the TFWP has been seen as "a breeding ground for migrant worker exploitation and labour market distortion" (Mertins-Kirkwood 2015, 151). The numbers of temporary foreign workers increased rapidly in the early years of the Harper government, from 55,059 in 2006 to a high of 107,921 workers in 2015, even while unemployment levels in Canada were rising and wages were falling, partly because of the 2008–10 recession. Because of controversy associated with this programme, there was a rapid course correction, and the government temporarily suspended the food industry's participation in the TFWP. In 2016, the numbers of workers coming in under the TFWP declined to 51,170 (Government of Canada 2016b).

In contrast, the numbers of permits granted under the IMP continued to increase gradually throughout the 2000s, from 92,927 in 2007 to 288,325 in 2017 (Government of Canada 2016b). The number of new international treaties that create new pathways for migrant workers has expanded dramatically over the past 15 years, largely because of the Harper government's interest in expanding such agreements. According to Mertins-Kirkwood, the Harper government "deliberately and systematically" (2015, 154) expanded the IMP and the pathways associated with it. The inclusion of mobility provisions in international trade agreements is of particular concern for understanding the changes in Latin American mobility, since so many of Canada's trade agreements include Latin American countries, including NAFTA as well as bilateral agreements with Chile, Costa Rica, Peru, Panama, Colombia, and Honduras in addition to the Comprehensive and Progressive Agreement for Trans-Pacific Partnership, which include Chile, Mexico, and Peru. Canada has also been attempting to negotiate a free trade agreement with MERCOSUR and the Pacific Alliance (discussed in the conclusion to this volume). The new agreements also expanded the types of business owners and professions included (Mertins-Kirkwood 2015, 155). The total number of IMP participants admitted to Canada under these trade agreements has increased substantially from 17,367 in 2007 to 36,673 in 2016. The largest number of these came under NAFTA (22,645 in 2016, up from 13,256 in 2007), but the number coming in under other agreements has also increased from 605 in 2007 to 1,046 in 2016 (IRCC 2017).[4]

Beyond the immediate short-term impact of FTAs, scholars argue that there is a long-term association between migration and the signing of trade agreements. Orefice argues that in addition to specific migration-related provisions in the FTAs, the signing of a trade agreement may affect bilateral migration flows by increasing familiarity and improving diplomatic relations among signatory countries. In her analysis of the 96 FTAs mapped by the WTO, she shows that migration increases by, on average, 26.7 per cent among partner countries when an FTA is signed, and by 33 per cent if that FTA include visa and asylum provisions (2015, 312–14).

Overall, changes in the Canadian migration system have had diverse effects on Latin American migrants. These changes have opened up new channels for mobility but have also maintained and deepened inequalities in treatment and access for Latin American migrants. In general, Latin Americans are less likely to gain access to permanent migration than migrants from some other regions because they are less likely to have the skills, fluency in English and/or French, and training to qualify them for the preferred migration categories. As well, Latin Americans fleeing violence in their homelands have trouble making successful asylum claims in Canada, partly because people fleeing violence caused by non-state actors like drug cartels are often not accepted as legitimate refugees, instead being classified as economic migrants, and partly because of the impact of the Safe Third Country Agreement (discussed below). At the same time, lower-skill Latin American migrants have gained increased access to the Canadian labour force but mostly on a temporary basis as an easily exploitable workforce under highly vulnerable conditions.[5] However, some higher-skill, higher-class Latin Americans may indeed have increased possibilities of attaining permanent status through such mechanisms as the mobility provisions in FTAs.

New Bordering Practices

The migration of Latin Americans to Canada especially within humanitarian streams has been adversely impacted by efforts of the Canadian government to police its borders and restrict asylum seekers. Two measures, the Safe Third Country Agreement (STCA) and the imposition of a visa requirement on Mexicans to Canada illustrate this.

The signing of a Safe Third Country Agreement between Canada and the United States in 2002 sought to limit the number of the asylum claimants at Canadian land border ports of entry and came into effect in December 2004. This was part of a broader "Smart Border Accord" signed by the two governments in the wake of the 11 September 2001 Al Qaeda attacks. In addition to commitments to pursue increased information sharing, preclearance, shared border facilities, common standards, etc., the Smart Border agreement also included measures to share information on asylum seekers to "identify potential security and criminality threats and expose 'forum shoppers' who seek asylum in both systems" (Government of Canada 2002).

Article 4(1) of the STCA states: "[t]he Party of the country of last presence shall examine, in accordance with its refugee status determination system, the refugee status claim of any person who arrives at a land border port of entry ... and makes a refugee status claim." (Government of Canada 2002). While the agreement formally applies to both countries, it effectively acts to limit the number of people able to make refugee claims in Canada, since would-be claimants are far more likely to travel by land through the United States to Canada to make a claim than in the opposite direction. The United States agreed to this request, despite the fact that it increases the numbers of individuals making refugee claims in the United States, in exchange for Canada's cooperation with other security measures (Cooper 2018).

In 2000 over one-third of Canadian refugee claims were filed from the United States (Clarkson 2008, 378). The agreement means that Canada must turn away any individual who arrives at an official point of entry from the United States, preventing him or her from making a refugee claim in Canada, since the United States is considered a "safe country" to make such a claim. The Canadian government defended this measure based on the argument that the United States had similar policies to Canada regarding refugee rights (even though Canada has been considerably more generous in accepting such claims). The agreement had a dramatic impact on the number of refugee claims filed in Canada: In 2005 Canada received just over 4,000 claims at border points of entry, down from approximately 8,900 claims filed in 2004. The agreement likely has a disproportionate impact on would-be claimants from the Americas, since they are the ones most likely to have been able to make a claim at a land border after passing through the United States. According to the Canadian Council for Refugees, Colombians were most strongly affected, since the number of claims submitted by asylum-seekers from that country declined by more than half between January and November 2005. Most Colombians had entered the United States first before travelling to the Canadian border to apply for asylum (Cowger 2017).

This agreement has been highly controversial. Amnesty International Canada and the Canadian Council for Refugees have engaged in a protracted legal campaign to have STCA rescinded and, in the interim, to immediately suspend the agreement. In a brief submitted to the then Liberal Immigration Minister Ahmed Hussen they argue that the application of the agreement "poses a significant threat to refugees in North America, by returning asylum-seekers to US authorities despite well-documented failings in the US refugee protection system. In so doing, Canadian practice currently violates both international and domestic norms" (Amnesty International Canada and Canadian Council for Refugees 2017: 3). Based on 2017 interviews with refugee claimants who had irregularly crossed the border into Canada, Amnesty International argues that most of these claimants have left the United States because of perceived failings in the US asylum system, xenophobic treatment, and fear that the human rights situation in the United States might deteriorate under the Trump administration (Amnesty International Canada and Canadian Council for Refugees 2017). Amnesty International Canada, the Canadian Council for Refugees and the Canadian Council of Churches have joined a Salvadoran woman, "E.," in asking the Federal Court to strike down the Safe Third Country Agreement and allow her to make a refugee claim in Canada. E. fled her country with her daughters after a decade of being targeted by a gang, including, most recently, death threats. She believes that she might not be protected if forced to make her refugee claim in the United States, rather than Canada. In July 2020 a federal court justice ruled that the STCA violates the Canadian Charter of Rights and Freedoms and gave the government six months to respond; the Liberal government has appealed this decision (Tunney 2020). During the pandemic, the government banned all asylum claims at unauthorized locations on the land border, citing health concerns. The ban was lifted on 22 November 2021 (CBC 2021).

In another example of increased border control affecting Latin American mobility, in 2009 Canada's relations with its NAFTA partner were sorely tested when Canada decided to impose a visa requirement on all Mexicans visiting Canada. The Harper administration stated that the Mexican visa was a necessary measure, pointing to the fact that in the previous three years Mexico was the top source country for persons seeking refugee status in Canada. Months before the visa imposition the Conservative Minister of Citizenship and Immigration, Jason Kenney, pointed out that the majority of Mexican claims were rejected: "That would indicate to me that the vast majority – something like 90 per cent of these claimants – are actually trying to immigrate to Canada through the back door of the refugee system and I think that unacceptable. That's basically queue jumping" (CBC News 2009). Paloma Villegas argues that Kenney's public pronouncements in effect identified "Mexican refugee claimants as having the potential to 'cheat' the system and therefore act in irresponsible ways. This connection had a racialized component. It was claimants' nationality that made them suspect in the media. Put differently, Mexican claimants were identified specifically through their Mexicanness, generalizing all Mexicans as potentially suspect" (2013, 2207-2208). Thus, all Mexican migrants became a risk that needed to be managed.

The effect of the visa on Mexican mobility was quickly apparent. "The decision to impose a visa on Mexican citizens in 2009 dramatically decreased the number of Mexicans who travelled to Canada, falling from 124,000 in 2010, from 271,000 in 2008" (Van Haren and Masferrer 2019). Calls for a change in Canadian policy came not just from Mexico but also from the Canadian business community and other commentators. A report authored by Laura Dawson and commissioned by the Canadian Council of Chief Executives (CCCE) noted that spending by Mexican tourists to Canada had fallen from $365 million in 2008 to $200 million in 2012. "Canada has paid a high price for the visa in terms of reputation and lost revenue from tourists, students, and investors. Canadian airlines have had to eliminate or reduce planned routes, and it is virtually impossible for a Mexican to arrange to travel to Canada on short notice, whether for business purposes or to take advantage of Canada's competitive air fares to Asia" (Dawson 2014, 17). It further observed that Canada's "imposition of a complex and intrusive 'temporary' visa that has lasted for more than five years is perceived as an insult to Mexican leaders and has chilled relations with Canada" (Dawson 2014, 16; Blanchfield 2014). The visa requirement was only lifted in 2016 following a change in the Canadian government. Almost immediately the numbers of Mexicans travelling to Canada increased to 393,000 in 2017, a figure that surpassed the 2008 levels (Van Haren and Masferrer 2019).

Conclusions

Canada's future in the Americas cannot be addressed without taking into account the increasingly complex migration patterns that shape and are shaped by the shifting political and economic dynamics of a transnational hemispheric space. As this volume documents, trade liberalization, the Harper Strategy, and security concerns have profoundly altered the spaces that people traverse in new and diverse ways. In this chapter we described and traced how Latin American migration to Canada, initially small in numbers and dominated by individuals fleeing violence, grew and was increasingly characterized by diverse movements of people. These latter paths included the entry of economic migrants reflecting the changes in the selection policies of the Canadian government. Increasingly, Latin Americans were entering in all categories, not just humanitarian streams. More recent policy developments, such as the expansion of the temporary worker programme and mobility provisions of trade agreements, mean many Latin Americans in Canada are admitted only on a temporary basis and their status is characterized by precarity. These developments have also been accompanied by recent measures that tighten the border, such as the Safe Third Country provision and the Mexican visa requirement, which act to restrict entry to Canada. We have argued that the rapidly changing nature of migration patterns and their complexity mean that conventional assumptions used to analyse Latin American migration, such as the familiar push-pull model and tropes of waves of migrants, need to be problematized, as they may not capture all new dynamics associated with such trends as circular migration and diaspora politics. Transnationalism may provide one new promising conceptual direction with its emphasis on linkages and connections across spaces.

NOTES

1 The Temporary Foreign Worker Program (TFWP) permits Canadian employers to hire foreign nationals as temporary workers to fill temporary labour and skill shortages when qualified Canadian citizens or permanent residents are not available.

2 Since September 2017, the Secretary of Homeland Security has announced plans to terminate TPS for six countries – El Salvador, Haiti, Honduras, Nepal, Nicaragua, and Sudan. TPS was extended for Somalia, South Sudan, Syria, and Yemen. Several lawsuits have been filed contesting the termination of TPS status (Congressional Research Service 2019).

3 The number of workers admitted as live-in caregivers reached a high of 41,707 people in 2009 and declined to 20,466 in 2015 (Government of Canada 2016b).

4 The Canada–US–Mexico Agreement (CUSMA), which was the outcome of the renegotiation of NAFTA that Trump demanded did not change the work categories under which temporary entry would be permitted, despite the concerns of the Canadian government and Canadian employers that these categories did not respond to changing workforce demands.

5 This vulnerability has become even more clear during the COVID-19 crisis, when Canadian farms became the site of some of the most severe outbreaks of the virus, with many cases occurring among migrant agricultural workers, including many Latin Americans.

REFERENCES

Amnesty International Canada and Canadian Council for Refugees. 2017. *Contesting the Designation of the US as a Safe Third Country.* Accessed 26 October 2021. https://ccrweb.ca/sites/ccrweb.ca/files/stca-submission -2017.pdf.

Armony, Victor. 2014. "Latin American Communities in Canada: Trends in Diversity and Integration." *Canadian Ethnic Studies* 46, no. 3: 7–34.

Berdichewsky, Bernardo. 2007. *Latin American's Integration into Canadian Society in B.C.* Vancouver: Canadian Hispanic Congress.

Blanchfield, Mike. 2014 "Canadian CEOs Call on Tories to End Mexican Travel Visa ahead of Harper Visit." Canadian Press (10 February). Accessed 22 November 2021. https://www.therecord.com/news/canada/2014/02/10 /ceos-call-on-tories-to-end-mexican-travel-visa-ahead-of-harper-visit.html.

Castles, Stephen, Hein De Haas, and Mark J. Miller. 2014. *The Age of Migration. Fifth Edition.* New York: Guilford Press.

CBC News. 2009. "Mexicans Trying to Immigrate 'through the Back Door,' Says Kenney." (15 April). Accessed 22 November 2021. https://www.cbc .ca/news/canada/calgary/mexicans-trying-to-immigrate-through-the -back-door-says-kenney-1.792456.

– 2021. "Quebec's Roxham Road Reopens to Asylum Seekers after Pandemic Ban." (22 November). Accessed 22 November 2021. https://www.cbc.ca /news/canada/montreal/roxham-road-reopen-1.6257868.

Clarkson, Stephen. 2008. *Does North America Exist? Governing the Continent after NAFTA and 9/11.* Toronto: University of Toronto Press.

Congressional Research Service. 2019. "Temporary Protected Status: Overview and Current Issues." Accessed 26 October 2021. https://fas.org/sgp/crs/homesec/RS20844.pdf.

Cooper, Celine. 2008. "A 'Safe Country' Dilemma for Canada." *Open Canada.* Accessed 29 November 2021. https://opencanada.org/safe-country -dilemma-canada/.

Cowger, Sela. 2017. "Uptick in Northern Border Crossings Places Canada-U.S. Safe Third Country Agreement under Pressure." *Migration Information Source* (26 April). Accessed 29 November 2021. https://www.migration policy.org/article/uptick-northern-border-crossings-places-canada-us-safe -third-country-agreement-under.

Dawson, Laura. 2014. "Canada's Trade with Mexico: Where We've Been, Where We're Going and Why It Matters." Canadian Council of Chief Executives, February. Accessed 26 October 2021. https://thebusinesscouncil.ca/app /uploads/2014/02/Canadas-trade-with-Mexico.-Where-weve-been-where -were-going-and-why-it-matters-February-2014.pdf.

Fischer, Lynette. 2017 "Trading on Mobility." *Briarpatch Magazine* 46, no. 1: 11–13.

García, María Cristina. 2006. *Seeking Refuge: Central American Migration to Mexico, the United States, and Canada.* Berkeley: University of California Press.

Ginieniewicz, Jorge, and Kwame McKenzie. 2013. "Mental Health of Latin Americans in Canada: A Literature Review." *International Journal of Social Psychiatry* 60, no. 3: 263–73.

Goldring, Luin. 2006. "Latin American Transnationalism in Canada: Does It Exist, What Forms Does It Take, and Where Is It Going?" In *Transnational Identities and Practices in Canada,* Edited by Lloyd Wong and Vic Satzewich, 180–201. Vancouver: UBC Press.

Goldring, Luin, Susan J. Henders, and Peter Vandergeest, 2003. "The Politics of Transnational Ties: Implications for Policy, Research, and Communities." YCAR-CERLAC Workshop Report. Accessed 26 October 2021. https:// www.yorku.ca/cohesion/LARG/PDF/Goldring_03_thepolitics.pdf.

Goldring, Luin, and Patricia Landolt. 2009. "Immigrant Political Socialization as Bridging and Boundary Work: Mapping the Multi-layered Incorporation of Latin American Immigrants in Toronto." *Ethnic and Racial Studies* 32, no. 7: 1226–74.

– 2011. "Caught in the Work-Citizenship Matrix: The Lasting Effects of Precarious Legal Status on Work for Toronto Immigrants." *Globalizations* 8, no. 3: 325–41.

– 2014. "Transnational Migration and the Reformulation of Analytical Categories: Unpacking Latin American Refugee Dynamics in Toronto." In *The Practice of Research on Migration and Mobilities,* edited by Liliana Rivera-Sánchez and Fernando Lozano-Ascencio, 103–27. New York: Springer. DOI:10.1007/978-3-319-02693-0_5.

Government of Canada. 2002. "Final Text of the Safe Third Country
    Agreement." Accessed 29 November 2021. https://www.canada.ca
    /en/immigration-refugees-citizenship/corporate/mandate/policies
    -operational-instructions-agreements/agreements/safe-third-country
    -agreement/final-text.html.
– 2016a. *Facts & Figures 2016: Immigration Overview – Permanent Residents –
    Annual IRCC Updates.* Accessed 26 October 2021. https://open.canada.ca
    /data/en/dataset/1d3963d6-eea9-4a4b-8e4a-5c7f2deb7f29.
– 2016b. *Facts & Figures 2016: Immigration Overview – Temporary Residents
    -Annual IRCC Updates.* Accessed 26 October 2021. https://open.canada.ca
    /data/en/dataset/6609320b-ac9e-4737-8e9c-304e6e843c17.
Hernández-Ramírez, Alejandro. 2019. "Mexican Youth in Canada: A Political
    Economy Analysis of Motivations for Immigration, Labour Market
    Integration, and Transnational Practices." PhD diss., Department of
    Sociology and Anthropology, Carleton University, 22 July.
Immigration, Refugees and Citizenship Canada (IRCC). 2016. *Annual Report to
    Parliament.* Accessed 26 October 2021. https://www.canada.ca/en
    /immigration-refugees-citizenship/corporate/publications-manuals
    /annual-report-parliament-immigration-2016.html#s2.
Landolt, Patricia, Luin Goldring, and Judith K. Bernhard. 2011. "Agenda
    Setting and Immigrant Politics: The Case of Latin Americans in Toronto."
    American Behavioral Scientist 55, no. 9: 1235–66.
Macdonald, Laura. 2018. "Canada Goes Global: Building Transnational
    Relations between Canada and the World, 1968–2017." *Canadian Foreign
    Policy Journal* 24, no. 3: 358–71.
Mata, Fernando B. 1985. "Latin American Immigration to Canada: Some
    Reflections on the Statistics." *Canadian Journal of Latin American and
    Caribbean Studies* 10, no. 20: 27–42.
Mertins-Kirkwood, Hadrian. 2015. "The Hidden Growth of Canada's Migrant
    Workforce." In *The Harper Record: 2008–2015,* edited by Teresa Healy and
    Stuart Trew, 149–58. Ottawa: Canadian Centre for Policy Alternatives.
Mignolo, Walter D. 2005. *The Idea of Latin America.* Malden, MA: Blackwell.
Ogelsby, J.C.M. 1976. *Gringos from the Far North: Essays in the History of
    Canadian-Latin American Relations, 1866–1968.* Toronto: Macmillan.
Orefice, Gianluca. 2015. "International Migration and Trade Agreements: The
    New Role of PTAs." *Canadian Journal of Economics* 48, no. 1: 310–34.
O'Reilly, Karen. 2016. "Migration Theories: A Critical Overview." In *The
    Routledge Handbook of Immigration and Refugee Studies,* edited by Anna
    Triandafyllidou, 25–33. Oxford: Routledge.
Prokofenko, Elena, and Feng Hou. 2018a. "How Temporary Were Canada's
    Temporary Foreign Workers?" Statistics Canada Analytical Studies Branch
    Research Paper Series, January.

‒ 2018b. "How Temporary Were Canada's Temporary Foreign Workers?" Statistics Canada, PowerPoint presentation. Accessed 26 October 2021. http://www.oecd.org/migration/forum-migration-statistics/3.Feng-Hou.pdf.

Riaño-Alcalá, Pilar, and Luin Goldring. 2014. "Upacking Refugee Community Transnational Organizing: The Challenges and Diverse Experiences of Colombians in Canada." *Refugee Survey Quarterly*, 1–28.

SICREMI. 2011. *International Migration in the Americas*. Washington: Organization of American States. Accessed 27 November 2021. https://www.oecd.org/migration/48423814.pdf.

Simmons, Alan B. 1993. "Latin American Migration to Canada: New Linkages in the Hemispheric Migration and Refugee Flow System." *International Journal* 48, no. 2: 282–309.

Stevenson, Brian. 2000. *Canada, Latin America, and the New Internationalism: A Foreign Policy Analysis, 1968–1999*. Montreal, Kingston: McGill-Queen's University Press.

Tunney, Catherine. 2020. "Canada's Asylum Agreement with the U.S. Infringes on Charter, Says Federal Court," CBC News (22 July). Accessed 26 October 2021. https://www.cbc.ca/news/politics/safe-third-country -agreement-court-1.5658785.

Van Haren, Ian, and Claudia Masferrer. 2019. "Mexican Migration to Canada: Temporary Worker Programs, Visa Imposition, and NAFTA Shape Flows." *Migration Information Source* (20 March). Accessed 27 November 2021. https://www.migrationpolicy.org/article/mexican-migration-canada.

Van Hear, Nicholas, Oliver Blackwell, and Katy Long. 2018. "Push-Pull Plus: Reconsidering the Drivers of Migration." *Journal of Ethnic and Migration Studies* 44, no. 6: 927–44.

Veronis, Luisa. 2007. "Strategic Spatial Essentialism: Latin Americans' Real and Imagined Geographies of Belonging in Toronto." *Social and Cultural Geography* 8, no. 3: 455–73.

‒ 2010. "Immigrant Participation in the Transnational Era: Latin American Experiences with Collective Organizations in Toronto." *International Journal of Migration and Integration* 11: 173–92.

Villegas, Paloma. 2013. "Assembling a Visa Requirement against the Mexican 'Wave': Migrant Illegalization, Policy and Affective 'Crisis' in Canada." In *Ethnic and Racial Studies* 36, no. 12: 2200–19.

7 "Advocacy Chill" or "Sunny Ways"? Harper and Trudeau and the Campaign to Reshape the Role of Civil Society Organizations[1]

KALOWATIE DEONANDAN AND TOVELI SCHMULAND

When Justin Trudeau and the Liberal Party campaigned against the governing Conservative administration of Stephen Harper in 2015, they appealed to the electorate with an array of progressive policies, including a strong de-emphasis on Harper's explicit pro-corporate agenda. Their platform was particularly appealing to civil society organizations (CSOs)[2], especially those social justice non-governmental organizations (NGOs) working on development issues in the Global South, including Latin America. To these groups, the Liberal programme was a welcome change from the open hostility and "advocacy chill" (ONN 2016) they faced during the Conservative era. Then, they were ostracized and sidelined through an array of policy manoeuvres, including tax audits, funding cuts, and other forms of bureaucratic control. What rendered them particularly vulnerable to the Conservative agenda is the fact that they rely heavily on federal funding to fulfil their mission.

In his efforts to oust Harper, Trudeau built his campaign around the promise to do politics differently and rallied support under a banner of "sunny ways," implying a more inclusive and cordial relationship with the different sectors of society, including the non-profit. As the Liberals have now ended one term in office and are into their second, it is instructive to inquire whether there are indeed differences between their approach to CSOs and that of their Conservative predecessor. While the Conservatives targeted almost all progressive groups (both those with a national and those with an international focus and those working on issues from poverty to anti-mining), the focus of this analysis is largely, though not exclusively so, on CSOs whose work is relevant to Latin America and which relates to the conduct of Canadian mining companies in the hemisphere.

The contention of this analysis is that there are indeed differences between the two governments when it comes to their approaches to progressive civil society organizations. The Liberals have consistently maintained a more tolerant and sympathetic attitude towards them and have employed a more inclusive rhetoric than their predecessor. At times, they have even implemented policies in keeping with the preferences of these organizations. This is significantly different from the Conservatives who sought to delegitimize and marginalize such groups in a coordinated campaign to neutralize them. However, in terms of policy goals, the two governments share a common agenda. Both sought to advance and protect Canada's economic interests, including its mining investments, both at home and abroad, and to ensure that CSOs play an instrumental role in this process. As a result, when major private-sector interests, such as the mining industry, have been targeted for criticism by these groups, there is evidence of convergence in the policy responses between the two administrations.

For Stephen Harper, promoting his policy agenda followed a more linear path, relatively speaking. Unapologetically neoliberal, he was a strong advocate of the free market, fiscal restraint, and private initiatives, and not surprisingly, he privileged industry and elite voices in policy development. While he opposed a socially activist state, he nevertheless was willing to invoke its powers to weaken, delegitimize, harass, and even criminalize opponents and critics, namely, progressive civil society groups. Meanwhile, for Trudeau, while he shares with Harper a commitment to promoting the interests of Canada's private sector, he also embraces a more socially progressive agenda, one that involves a more activist state, increased social spending, and the promotion of diversity and inclusivity. His challenge is to balance these seemingly competing agendas.

That Trudeau and Harper share common objectives with respect to Canada's economic interests is not surprising. Canada is a capitalist state, and its success demands that it continues to grow and expand. Consequently, it needs to ensure that it is competitive globally, that it exploits its competitive advantages, and that it safeguards the interests of its private sector – even in the face of opposition from the wider society. Where Harper wielded a blunt axe in his efforts to silence critics of his economic strategy, Trudeau is attempting to navigate a fine balance by simultaneously courting both the corporate sector and civil society groups. Maintaining the support of the latter is crucial; it was especially so as Trudeau's second electoral contest loomed, and it continues to be so as his government's power rests on a minority parliament. Stephen Brown (2017) summarized Trudeau's approach as "Harper Lite."

In analysing the strategy that each government adopted vis à vis CSO society, this chapter will focus on the policy domains that proved contentious for each government. Five issue areas will be addressed: the Charities legislation, the consultation process with CSOs, funding, the establishment of the office of the Canadian Ombudsperson for Responsible Enterprise (CORE), and the Gender Equality and the Feminist International Assistance Policy (FIAP). In demonstrating its claims, the remainder of the paper will be divided into five sections. The first summarizes the methodology, the second outlines the theoretical framework, the third discusses the approach of the Harper administration towards NGOs, the fourth analyses the strategies of the Trudeau government with respect to these civil society groups, and the final section offers concluding remarks.

Methodology

This chapter relies on qualitative research based on primary materials, including phone interviews and email communications with three Canadian NGOs (KAIROS, MATCH, and MiningWatch Canada) working in overseas development in Latin America and with a strong interest in the government's policies on resource development. As well, it draws from publications by these and other NGOs and from statements of relevant government leaders and organizations. Peer-reviewed publications were utilized for the theoretical framework and for the background information.

Theoretical Approach

To understand Trudeau's and Harper's approaches to Canadian civil society groups, the theoretical lens that guides this analysis borrows from elements of the framework developed by Peck and Tickell (2002) and expanded upon by Brenner, Peck, and Theodore (2010). Their schema revolves around the concept of neoliberalization, which, in short, refers to the strategies devised by free-market advocates to respond to the failures of neoliberalism. While neoliberalization in the authors' conceptualization involves several components, that which is applicable to the present context is the process of regulatory transformation. This entails both "destructive" and "constructive" moments. The former encompasses the standard neoliberal policy package of privatization, deregulation, and the retreat of the state, that is the "roll-back" ("roll-back neoliberalism") or declawing of regulatory standards that hinder market operations in a wide array of areas, including the financial, social, and environmental. Conversely, the latter, the "constructive" moments of the regulatory transformations under neoliberalization demands new regulatory experiments ("roll-out neo-liberalism"). Important in this phase is the role of the state. Far from retreating, it is crucial in building this new architecture of regulations to reconstitute and bolster neoliberalism and its own authority. As Peck and Tickell (2002, 384) elaborate: "In the course of this shift [to roll-out neoliberalism], the agenda has gradually moved from one preoccupied with the active *destruction and discreditation* [*sic*] of Keynesian-welfarist and social collectivist institutions (broadly defined) to one focused on the purposeful *construction and consolidation* [*sic*] of neoliberalized state forms, modes of governance and regulatory relations."

Hence, while the prevailing conceptualization of the state under neo-liberalism is that it is in retreat, in fact, as Catherine Kingfisher (2002, 8) stresses in her writings on the role of the neoliberal state in the feminiz-ation of poverty, the state must be kept "clearly in view." Its interven-tion, however, is selective, as is manifested in "the striking co-existence of technocratic economic management and invasive social policies" (Peck and Tickell 2002, 389). The "manipulation of interest rates, the maintenance of noninflationary growth, and the extension of the 'rule' of free trade abroad and flexible labour markets at home" are all hall-marks of this form of economic management (389). At the same time, the activist state is also very much in evidence when it comes to social policies. As Peck and Tickell (2002) observe: "[A] deeply interventionist agenda" prevails on 'social' issues like crime, immigration … welfare reform … [and] community regeneration. In these latter spheres, in par-ticular, new technologies of government are being designed and rolled out, new discourses of 'reform' are being constructed (often around new policy objectives…) new institutions and modes of delivery are being fashioned, and new social subjectivities are being fostered" (389). The neoliberalization framework outlined helps to structure our understanding of both the Conservative and Liberal approaches to civil society groups. As will be discussed below, both administrations were motivated by the same objectives, to promote and protect Canada's national and overseas investments and to neutralize or marginalize domestic criticisms of their government. To achieve this, both adopted "invasive" social policies and regulatory reforms to discipline their civil society critics. For its part, the Harper government wielded very direct instruments of control. These included promoting public–private part-nerships between NGOs and mining companies; legislating new reg-ulatory standards, such as the Political Activities Audit Program and the Not-for-Profit Corporations Act, which were designed ostensibly to ensure transparency and good governance but which were in essence forms of administrative harassment; and defunding organizations critical of it. This is not to suggest that civil society groups were not mobilized by the Harper government to advance the Conservative agenda. Indeed, several organizations, such as those funded to work with Canadian mining companies in Latin America and Africa in corporate social responsibility (CSR) programmes, served an instrumental role for the administration by advancing the interests of Canada's mining sector. Others were defunded towards this same end.

Trudeau too embarked on a process of regulatory experimentation to bring NGOs more in line with his policy goals. Unlike Harper's strategy of outright alienation and punishment, his approach was to placate rather than alienate. He opted for inclusivity in rhetoric, and even occasionally in practice, and generally navigated a policy path that signalled progressivism, while concurrently guaranteeing corporate protections. He instituted a wave of consultations with NGOs, halted the Canada Revenue audits but did not initially repeal the legislation that permitted them, appointed a long-demanded Ombudsperson to oversee the conduct of Canadian mining firms abroad but deferred empowering it, and implemented a Feminist International Assistance Policy (FIAP) to support the advancement of women but which does so through market mechanisms.

What the aforementioned summary reveals is that despite their differences, what Harper and Trudeau shared is that they each sought to protect Canada's corporate sector while simultaneously maintaining the support of their respective political base, which come from different ends of the political spectrum (Brown 2016).

The Harper Era and Civil Society

Stephen Harper's policy towards Latin America had direct implications for his government's approach to NGOs working in the region. His Americas Strategy was articulated during his visit there in July 2007, when he announced that re-engagement with the region is a leading priority on his government's foreign-policy agenda. This was a dramatic shift from the government's inattention to the relationship for the previous decade. Canada's objectives in the region, he asserted, were threefold: to strengthen democracy, promote economic growth and investment, and enhance security (Gov't of Canada 2007). What underpinned this urge for renewed engagement with the Americas was the need to promote and protect Canada's growing financial interests in the region.

More specifically, the agenda was driven by the very substantive and steadily rising investments in Latin American by Canadian extractive firms. By 2016, according to Natural Resources Canada (NRCAN 2018), approximately 50 per cent of the world's publicly listed mining and exploration firms were in Canada and 54 per cent of these had investments, valued at $88 billion in Latin America and the Caribbean. At the same time, according to the Observatory of Mining Conflicts in Latin America (OCMAL 2020), the region was plagued with over 200 anti-mining conflicts, many of which involved Canadian-owned operations.

Meanwhile, at home in Canada, a growing number of development organizations working in the region were raising their voices in protest of both the conduct of the Canadian mining companies and the pro-mining policies of the Canadian government. Very swiftly, and harnessing the authority of various sectors of the state bureaucracy, the Harper government countered with an all-encompassing retaliatory strategy to delegitimize and discipline these groups. Four measures were of particular significance in this disciplinary process: the promoting of public–private partnerships, the implementation of the Political Activities Audit Program (PAAP), the de-funding of civil society critics, and the adoption of the Not-for-Profit Corporations Act (NFP Act).

Public–Private Partnerships – NGOs and Mining Companies

Telling of this pro-business orientation was the Harper government's decision to link the protection of Canada's mining interests in Latin America to its foreign-aid programme, and to mobilize some civil society groups to advance that goal. This policy was formalized in 2013 with the merging of Canada's official aid organization, the Canadian International Development Agency (CIDA) with the Department of Foreign Affairs and International Trade (DFAIT). The government justified the fusing of these two agencies by asserting a linkage between Canada's economic interests and its foreign-aid programme, making explicit what was already an implicit practice. In explaining the policy, the Minister for International Cooperation, Julian Fantino, stated that the primary goal of CIDA, the agency established initially to fight poverty overseas, was now "to make countries and people, trade and investment ready" (quoted in Gordon 2012). The façade of humanitarianism in overseas development assistance was all but eliminated as the pro-business priorities were pushed unambiguously to the forefront of the aid agenda.

The commercialization of aid policies became even more entrenched when CIDA announced funding for public–private partnerships (PPPs) between three Canadian NGOs operating in Africa and Latin America – Plan Canada, World University Services Canada, and World Vision – with three large mining firms, respectively, IAMGOLD, Rio Tinto, and

Barrick, to fund CSR programmes in these regions (Gov't of Canada 2011a). Under the agreement, the government provided $6.7 million over five and a half years while the mining companies, combined, contributed just under $2 million (Gov't of Canada 2011a).

Critics charged that far from promoting development, the actual aim of these PPPs was to sanitize the negative image of mining companies and quell the widespread resistance the latter face from host communities harmed by their operations (Schulman and Nieto 2011). The government was also denounced by groups such as War Child for subsidizing wealthy corporations, which earn billions in profits, with taxpayer dollars (Nutt 2012). Barrick Gold, for example, which was partnered with Plan Canada to work in Peru, is the world's largest gold-mining company, with reported revenues in 2017 of US$8.37 billion and net earnings of US$1.44 billion (Barrick 2018). It also operates some of the most controversial mining projects around the globe, including Pascua Lama in Chile and Porgera in Papua New Guinea. According to Stephen Brown (2016, 283–4), "the greatest beneficiaries of [CIDA's PPP program were] ... the mining companies themselves: public funds enhance their ability to operate in mining-affected communities," conferring legitimacy on their activities.

The policy not only led to criticisms of the government, but it also damaged the reputations of the NGO participants. Samantha Nutt, executive director of War Child Canada, a development organization that received CIDA funding, commented that the PPP paradigm established a dangerous precedent and undermined the credibility of civil society groups as it signalled that aid organizations are "no longer involved in aid delivery work, but are in the sub-contracting business, and forced to bid on aid" (Nutt 2012). Brown (2016) too noted the challenges posed for NGOs by the PPP model of aid delivery. Among other consequences, he claimed that it drove a wedge between those groups that participated and those that did not, it led to a loss of credibility for those that were involved in partnership arrangements, and it placed NGOs in competition with each other for government funding.

Political Activities Audit Program (PAAP)

Detrimental though their impact may have been for NGOs, their participation in PPPs represented the "carrot" approach by the Conservatives in their dealings with civil society organizations. The "stick" was yet to come. It took the form of bureaucratic harassment, which targeted those groups critical of Harper's policies, specifically his unbending support of the mining, oil, and gas sectors, and notably the pipeline projects.

The most aggressive manifestation of the "stick" strategy was the $13.4 million Political Activities Audit Program (PAAP), which began in 2012 and aimed to neutralize critics (Beeby 2016a). Essentially, the programme directed the Canada Revenue Agency (CRA) to conduct close to sixty audits of CSOs (Beeby 2014) to determine the legality of their activities; the objective was to determine whether these were primarily charitable or whether they had veered beyond allowable limits into the political sphere. In Canada, charities are governed by the Income Tax Act administered by the CRA, and the agency's Charities Directorate oversees the application of the Act with respect to charitable organizations. One of the Act's stipulations is that registered charities are restricted to utilizing only 10 per cent of their resources for political activities and are prohibited from undertaking partisan political activities.

Triggering the decision to launch the audit programme was the government's anger over the efforts by environmental groups to halt the pipeline projects in Western Canada. According to social media scholar Gareth Kirkby, many of these audits were instigated by Ethical Oil, a lobby group with close links to the Harper government and the oil industry (quoted in Goar 2014). Ethical Oil had launched a grievance with the CRA, charging that environmental groups, including Environmental Defence, Tides Canada Foundation, and the David Suzuki Foundation, had violated the 10 per cent spending cap allowed to charities for political activities (CRA Charity Audits, n.d) and that many of these groups were also involved in money-laundering activities (issuing charitable tax receipts on behalf of third -party organizations that were not qualified to do so) (O'Neil 2012). These allegations provided the rationale for the Conservatives to direct the CRA to implement the targeted audits, a practice not followed generally by the agency. As Goar (2014) noted, "Unlike the random audits it [the CRA] has always conducted … the new ones [were] susceptible to external direction, compromising the fairness and professionalism in which the tax department has always taken pride." As with other Conservative tactics against its NGO critics, the audit strategy seemed designed to drain energy and resources from these groups to restrict their capacity for mobilization against the government. The Toronto-based group Environmental Defence complained that it was forced to spend over $200,000 in legal fees to deal with the audit (Beeby 2018b).

Defunding NGO Critics

Prior to the PAAP, the government had already begun to rein in its critics within the NGO sector, and one of the most effective strategies was depriving them of their long-standing governmental financial support. The first organization targeted for defunding, and which was also subsequently the focus of a CRA audit, was KAIROS: Canadian Ecumenical Justice Initiatives. Among CSOs and their supporters, the suspicion was that these disciplinary measures were directly linked to KAIROS's condemnations of the actions of Canadian mining firms at home and abroad and of the government's focus on pushing extractive development. KAIROS was active in drawing attention to the damages inflicted on communities that host mining operations; it had called on the government to support Bill C-300 which sought to hold Canadian mining companies operating overseas accountable for their actions; and it had denounced the Alberta tar sands projects and had demanded the government cease further approvals of such proposals (KAIROS, n.d.). As well, it had lobbied for Canada to become a signatory to the United Nations Declaration on the Rights of Indigenous Peoples (UNDRIP). Accession to UNDRIP would impose specific obligations on the government in terms of its relationship with Indigenous peoples and could affect companies' gaining approval for extractive operations, such as the oil sands project.

The defunding of KAIROS came in 2009 with similar disciplinary measures taken against close to a dozen other organizations the following year. Among them were MATCH International, the International Planned Parenthood Federation (IPPF), Alternatives, Development and Peace, and the Canadian Council for International Cooperation, all of which had criticized the government on different issues. As Barry-Shaw and Jay (2012) observed, "For the targeted organizations, the loss of government funding meant between 40 per cent and 75 per cent of their annual budgets disappeared overnight. The cuts exacted a heavy toll: overseas programmes were shut down, offices were closed, staff positions were eliminated, and properties were liquidated." Through these regulatory interventions the Harper government effectively defanged its civil society critics and cleared the way for advancing a corporate-friendly agenda, one specifically beneficial to Canadian extractive firms.

Not-for-Profit Corporations Act (NFP Act)

Complementing the above PAAP strategy was the passage of the Canada Not-for-Profit Corporations Act (NFP Act), which also aimed at neutralizing the government's civil society critics. The bill came into force in 2011 and replaced the old Canada Corporations Act. The new law required that non-profits, which were registered under the old statute, reapply to maintain their non-profit status and meet the over ninety requirements included in the regulations (Government of Canada 2011b). The aim of the Act was ostensibly to modernize the rules governing the non-profit sector. However, while this updating may have been necessary as NGOs themselves admit, it nevertheless imposed an onerous burden on many as they were compelled to invest a significant amount of time and resources to meeting the new criteria to maintain their charitable status and thus their funding. As was likely part of the government's intent, this detracted from the ability of the affected groups to focus on their organizational goals, as they were preoccupied with, for example, re-writing bylaws; clarifying the roles, duties, and powers of their directors; and filing new documentation updating their operational processes, etc.

Furthermore, by being subjected to these new conditionalities "they [were] inevitably drawn into supporting and even spreading many aspects of the dominant ... [neoliberal] agenda," as Wallace (2004) notes in her writings on international NGOs. There are indications that the Act was designed to render non-profits more like corporations when it comes to governance and administration. As one legal expert on charities and a supporter of the legislation explained, "It provides modern governance principles, standards and machinery – similar to the Canada Business Corporations Act" (Bridge, n.d). This corporatizing agenda was confirmed by the then Minister of State (Small Business and Tourism) Diane Ablonczy, who stated that the act was designed to "promote accountability, transparency and good *corporate* governance for the not for profit [*sic*] sector" (quoted in Industry Canada 2008, emphasis added). While advocating accountability, transparency, and good governance is tantamount to a motherhood statement, there are nevertheless critics who see these as codes or catchwords of neoliberalism. Allison Ayers (2006), writing about the democratization processes in the Global South and the push by international financial institutions, such as the World Bank, for best practices in terms of "transparency" and "good governance," points out that these are often buzzwords designed to restructure or "reengineer" civil society to advance private economic interests.

The punitive measures implemented by Harper's Conservatives ushered in for the target groups a climate of apprehension and anxiety, referred to by some as an "advocacy chill" (ONN 2016). For Canadian foreign policy scholar Stephen Brown, the strategy was designed to play to Harper's base – an important component of which was big business. Harper's successor, Justin Trudeau, campaigned on a platform of doing politics differently, both at home and abroad. The following section assesses the extent to which he adhered to this commitment, especially when it came to the treatment of CSOs, particularly those who were the target of Harper's disciplinary tactics.

Trudeau and Civil Society: Continuity and Change

Trudeau and the PAAP

During the 2015 electoral campaign, the Trudeau Liberals launched their thirty-two-point policy platform dubbed "Real Change" (The Liberal Party 2015), which outlined a more socially oriented programme than that of their predecessor. Prominent among their more progressive policies was their call for the empowerment of CSOs. Responding directly to the Conservative attack on the sector, the Liberals vowed that they would reverse this trend: "We will allow charities to do their work on behalf of Canadians free from political harassment, and will modernize the rules governing the charitable and not-for-profit sectors. This will include clarifying the rules governing "political activity," with an understanding that charities make an important contribution to public debate and public policy. A new legislative framework to strengthen the sector will emerge from this process" (The Liberal Party 2015, 34). This commitment raised the expectations of the groups targeted for neutralization by Harper-era policies that a more hopeful phase would be inaugurated with a Liberal victory. The record thus far, however, has been mixed as the Liberal government has been slow to act on several fronts in areas of crucial interest to civil society groups.

One particular concern was the ongoing CRA audits which non-profits had hoped would end once the Liberals took office. They were sorely disappointed, however, as the new government's stance proved ambiguous. In January 2016 the Minister of National Revenue, Diane Lebouthillier, announced that she would allow the twenty-four audits that were in progress to continue, would not intervene to halt the deregistration underway for five groups, but that she would cancel the six audits that had not yet begun (Beeby 2016). Her rationale was that the "independence of the Charity Directorate's oversight role for charities must be protected" (quoted in Government of Canada 2016). In essence, she invoked bureaucratic requirements as cover to avoid taking a decision on her predecessor's actions.

One of the charities targeted by the audits was Canada Without Poverty (CWP). While this group is not involved in overseas aid work, the actions taken against it had implications for civil society groups generally, including those involved in international development. As such, the fate of CWP was closely watched by the sector.

CWP had been entangled in the audit process since November 2011 and was scheduled for deregistration due to the CRA's determination that it violated the 10 per cent limit on political activities (98.5 per cent of its activities were deemed political by the audit) (Beeby 2016b). This impending declassification constrained the group's fundraising ability and threatened its very survival since it could not issue charitable receipts to donors. In response, CWP launched a Charter challenge against the legislation in the Ontario Superior Court in August 2016. It argued that the 10 per cent limit "infringed its right to freedom of expression" guaranteed by the Charter of Rights and Freedoms and was an "arbitrary" line that "limited its ability to engage in public policy advocacy, which is its central purpose" (Simard and Kudhail 2018). The Liberal government countered that the case was really about money and not principles: "The true nature of its [CWP's] application is a demand for financial support from the state ... Freedom of expression does not require the government to subsidize Canada Without Poverty's activities, and this application should therefore be dismissed" (quoted in Beeny 2018a), and it demanded reimbursement for legal costs incurred in the case. In its ruling handed down in July 2018, the court sided with CWP and lifted the 10 per cent limit.

The government's actions in this case and its response to this decision surprised and disappointed many. Approximately one month after the court's ruling, the Liberals announced they would appeal the decision and, in a joint statement, the Ministers of National Revenue and of Finance elaborated on the reasons why: "[T]the Government of Canada has identified significant errors of law [on which the judgment was based] and has served notice that it will be appealing the decision to address the uncertainty created by it, and to seek clarification on important issues of constitutional and charity law. The resolution of these legal issues, while necessary, will not change the policy direction the Government intends to take with respect to the removal of quantitative limits on political activities" (Gov't Canada 2018). The statement seems to suggest that the government was trying to have it both ways. That is, it was claiming to support charities and willing to give them operational flexibility by lifting quantitative restrictions, yet it was challenging a ruling that removed those limits. The response of NGOs conveyed both their unease with this stance as well as their reluctance to criticize the government outright given that they had just endured the years-long open hostility of the Harper era. KAIROS's executive director, Jennifer Henry (2018b), stated, "we are concerned that the government has appealed even while signalling a policy change." Tim Gray of Environmental Defence responded with an even more conciliatory stance:

"We do not support the decision to appeal the ruling but recognize that the federal government is making moves to address the issues raised in it" (quoted Beeby 2018b).

Disturbing to civil society groups too was the government's lack of response to the recommendations of a blue-ribbon panel established by Minister Lebouthillier in September 2016, soon after the above court challenge had been launched by the CWP. The mandate of the panel was to study the regulatory context for charities and to make recommendations for improvements. The panel consisted of five prominent individuals, including a charities lawyer and the CEO of the David Suzuki Foundation; the latter group was also among those in Harper's crosshairs. In March 2017, less than a year after its formation, the panel issued its report with four major recommendations. One of these was that "any reference to non-partisan 'political activities' be deleted from the Income Tax Act "to explicitly allow charities to fully engage, without limitation, in non-partisan public policy dialogue and development, provided that it is subordinate to and furthers their charitable purposes" (Government of Canada 2017c). Another called for the modernization of the legislation governing charities and that a more "inclusive list of acceptable charitable purposes [be adopted] to reflect current social and environmental issues and approaches" (Government of Canada 2017c).

In response to the report, the Minister called for the suspension, but not cancellation, of the remaining twelve audits that were still underway, and she pledged to review the panel's recommendations and make a more fulsome response sometime in 2018 (Beeby 2018a). She also promised that the government will "amend the Income Tax Act to implement changes consistent with recommendation no. 3 ... and that these will allow charities to pursue their charitable purposes by engaging in non-partisan political activities and in the development of public policy" (Government of Canada 2018).

Governmental action on the report, however, was slow in coming. In January 2018, the Liberals established a Special Senate Committee on the Charitable Sector to study the government's relationship with the non-profit sector, but the committee did not begin meeting until eight months later. The delay led some CSOs to fear a "new chill" was setting in and that the Liberals "will not act on [their] ... promise to protect charities" (quoted in Beeby 2018a). The fear finally abated in December 2018, when the government amended the Income Tax Act under Bill C-86 and removed all references to political activities and charities (Beeby 2019). In place is now a *Guide* that informs charities on "public policy dialogue and development" (Government of Canada 2019a).

For CWA, a particularly important piece of news was the government's announcement that it was dropping its appeal of the court ruling against the group. It can be argued that the new legislation had rendered redundant the charges against the organization given the absence of reference to political activities in it.

These measures that the Liberals adopted, albeit after a lengthy delay, won them high praise from civil society groups, especially CWA. Leila Farha, Head of the organization, stated: "The decision to let Justice Morgan's decision [ie the initial ruling] stand is a huge victory for democracy in Canada ... The government has done the right thing twice. First they made the legislative changes recommended by the government's appointed panel and ordered by Judge Morgan, and now they have properly decided to withdraw their appeal ... This decision puts Canada in the lead among common law countries and will have a positive effect not only in Canada, but worldwide" (quoted in Beeby 2019).

What accounts for the delayed response by the Liberals in cancelling the audits and implementing the new legislation? It is not inconceivable that they had wanted to maintain the same political tools in their arsenal as the Conservatives to deal with those CSOs that might mobilize against them in the future. Preserving the 10 per cent limit would have given them that option. The eventual shift in their position can perhaps be attributed to their realization that this decision was more in line with their ideological stance and was commensurate with their electoral promise to protect charities. More importantly, they likely calculated that reneging on this electoral assurance might prove costly in the then impending federal elections (in 2019), given that the Party was sliding downwards in the polls. It was, therefore, imperative for them to preserve the loyalty of the progressive sector and prove to their political base that they can be trusted to abide by their promises. From the response of the CWA leader, it would seem that the backing of the progressive social groups was preserved with this move.

Consultation

One of the areas where Trudeau's government differs considerably from Harper's is in the former's penchant for consultation with civil society groups and its willingness to provide them with access to government officials. However, consultations do not commit the government to specific policy outputs, but it does convey the image of a highly engaged administration and can confer legitimacy on its actions. The Liberals have taken full advantage of available consultative mechanisms to keep the support of CSOs, especially those marginalized during the Harper era. They have also moved to institutionalize the consultation process by establishing a new agency to foster dialogue between the government and the charitable sector.

In May 2016, Marie-Claude Bibeau, Minister of International Development and La Francophonie, announced public consultations for the International Assistance Review, which aimed "to renew Canada's international assistance policy and funding framework ... to refocus Canada's international assistance on helping the poorest and most vulnerable people ... [to] determine Canada's approach internationally to supporting the 2030 Agenda on Sustainable Development" (Government of Canada 2017b). Over 300 consultations were held, a staggering 15,000 individuals and organizations participated, and there were more than 10,600 written submissions (Government of Canada 2017b) during this process. Among the participating groups were those the Harper government had defunded and discredited. According to Jennifer Henry of KAIROS, the first group targeted for the Harper cuts, the situation has "dramatically changed," and there is now a "very significant level of consultation," so much so that they are almost "overly consulted" (Henry 2018). Beth Woroniuk (2018) of MATCH Canada revealed that organizations refer to this particular period as the "summer of consultation." However, they both agreed that "consultation is always a good thing" (Henry 2018). Amnesty International (2017) too scored the government quite high on the consultation scale, noting that "There is clearly a new embrace of the language of human rights, including enthusiastic expressions of support for the Charter of Rights and Freedoms and Canada's international human rights obligations ... In many areas of government, there has been a notable new openness to engagement with civil society and the public. Many areas of law and policy have been the subject of consultations, parliamentary hearings and study."

Just prior to Trudeau's second federal electoral contest, the Liberals announced the creation of a permanent Advisory Committee on the Charitable Sector (ACCS) and established the agency in August 2019, thereby entrenching the consultation process. The new seventeen-member body (fourteen from the non-profit sector and three from government) is co-chaired by two prominent representatives from the charitable sector, Hilary Pearson, president and CEO of Philanthropic Foundations Canada, and Bruce MacDonald, president and CEO of Imagine Canada (Government of Canada 2019e). According to Minister Lebouthillier, the task of the new organization is to "provide an ongoing and substantive dialogue with the federal government around the policy and regulatory issues that affect the effectiveness

172 Kalowatie Deonandan and Toveli Schmuland

and sustainability of the charitable sector" (quoted in Government of Canada 2019).

While consultations and dialogue with the government are sign of major advances from the relationship with the previous Harper administration, critics argue that these represent form over substance, that they will not lead to significant policy changes but are simply "rebranding" exercises designed to cast the Liberal government in a more favourable light (Brown 2017a). Jen Moore (2018), then of MiningWatch Canada, an NGO that campaigns on behalf of communities affected by resource development, commented that, "despite the tremendous level of consultation ... what's actually coming out at the other end is not what people were calling for."

Further, it is tempting to suggest that the new advisory committee, the ACCS, was a politically motivated creation, as was the decision to cancel the audits. The new agency was established just as the Liberals' first mandate was ending and they were about to face the electorate again. Cognizant of the criticisms emanating from CSOs that had supported them but which had become apprehensive over concerns of a renewed advocacy chill, the Liberals likely acted to ensure the former remained on side during the upcoming contest. As noted earlier, polls were showing waning support for the Party. The ending of the aforementioned audits, the setting up of ACCS, and the creation of a new Ombudsperson's office, discussed below, were measures to reassure the civil society constituency and keep them in the Liberal fold.

The Canadian Ombudsperson for Responsible Enterprise (CORE)

One of the more contentious areas between the Liberal government and the progressive civil society sector involves the establishment of the office of the Canadian Ombudsperson or Responsible Enterprise (CORE), as discussed in the chapter in this volume by Paul Haslam. Not coincidentally, the powers and composition of this office provide one of the strongest indications of the Liberals' advancing their agenda via rollout neoliberalism. "Re-regulation" in this context is designed to simultaneously placate civil society and advance the interests of the business community, specifically in the resource sector. During the above-mentioned consultations, groups such as the Canadian Council for International Cooperation's (CCIC), Americas Policy Group, and others actively lobbied the government to implement reforms in resource governance[3] to address the gross charges of misconduct by Canadian mining firms overseas. In particular, they called on the government to establish an office that would: "receive complaints, thoroughly investigate harm caused or contributed to by Canadian companies, publicly report on its findings, make recommendations for remedy for affected communities, including the withdrawal of government support to companies involved in wrongdoing, and make recommendations to prevent future harm, including regarding Canadian law and policy reform" (CNCA 2020).

During their first electoral campaign in 2015, and in response to calls from groups such as the Canadian Network on Corporate Accountability (CNCA) for oversight of Canada's extractive industries, the Liberals pledged to "set up an independent ombudsman office to advise Canadian companies, consider complaints made against them, and investigate those complaints where it is deemed warranted" (quoted in CNCA 2016). Once in office, however, there were no further references to the pledge (Porras Ferreyra 2016; Moore 2018). Commented one observer in the *New York Times*, "Mr. Trudeau has been silent when it comes to one key issue for Latin Americans, an issue that has soiled Canada's image with thick layers of sludge: the reprehensible behavior of mining companies in the region" (Porras Ferreyra 2016).

It was not until January 2018 that the government finally addressed the issue and announced the creation of the office of the CORE. According to MiningWatch's then Latin American Program Coordinator Jen Moore (2018), the government only acted when, "it was facing the political need to do so." In other words, the decision was based on political calculation. As she elaborated, the Liberals were experiencing declining support due to their failure "to follow through on key promises [despite being] midway through their mandate ... The government was scrambling ... to show it was doing something which was when it started latching on to the ombudsman proposal" (Moore 2018).

While an extractive industry oversight body had existed under Harper, the Office of the Extractive Sector Corporate Social Responsibility (CSR) Counsellor it was unanimously dismissed by civil society groups for its lack of regulatory capacity and its ineffectiveness. Established in 2009 in the wake of mounting allegations of human rights abuses by Canadian mining firms abroad, the office's directive was to assist in resolving conflicts between mines and communities. However, for the NGOs working with the affected communities, the office lacked credibility and clout, as it possessed neither investigative powers nor enforcement authority to compel participation by firms in any dispute-resolution process (Bone 2015).

The Liberal creation CORE seemed, at first glance, to have more substantive powers than the Harper-era CSR Counsellor. Following CORE's creation, one civil society organization enthusiastically commented that Canada had become "the only country in the world to create an ombudsperson for responsible business conduct abroad" a move that "serves to put it on the map of jurisdictions evidencing a commitment to addressing corporate accountability" (ICAR 2018). CORE was also a more inclusive organization than Harper's CSR Counsellor as it incorporated strong representation from civil society groups. However, subsequent developments raised serious doubts regarding the credibility of the new body and whether it was in fact any different from its Conservative predecessor. One area of concern had to do with the governance of the organization.

According to its mandate, CORE is "to address complaints related to allegations of human rights abuses arising from a Canadian company's activity abroad ... and to undertake collaborative and independent fact-finding, make recommendations, monitor implementation of those recommendations, and report publicly throughout the process" (GAC n.d.). In terms of its enforcement capacity, the office "can recommend sanctions, which include the withdrawal of certain Government services, such as trade advocacy and future Export Development Canada (EDC) support, for companies found to be involved in wrongdoing" (GAC n.d.). Overseeing CORE is a new multi-stakeholder Advisory Body on Responsible Business Conduct composed of members from government, industry, and civil society. Among the latter are seven of the more prominent CSOs, many of which were targeted by Harper for their criticisms of both his administration and of the conduct of Canada's mining companies. Included in the ranks of the boards appointees are the Canadian Labour Congress, the United Steelworkers Union, the Canadian Network on Corporate Accountability (CNCA), Amnesty International, MiningWatch Canada, Development and Peace, and World Vision Canada. World Vision, as was noted earlier, had entered a CIDA-funded public–private partnership with Barrick Gold in Peru as part of the Harper-era strategy to rebrand the image of Canada's mining operations overseas. That it was included in the Liberal policy remodelling, along with groups strongly critical of the Conservative strategy, could be interpreted as evidence of Trudeau's more inclusive approach to doing politics. As for the powers of the Advisory Body, according to the International Corporate Accountability Roundtable (ICAR 2018), the group will develop "a set of guiding principles to assist the CORE in its operations as well as advising the Canadian government more generally on responsible business conduct by Canadian companies operating abroad."

While the creation of the advisory board and its membership structure may bring plaudits for the government, its leadership gave cause for concern. Heading the organization is Harvard Professor John Ruggie, who had served as the United Nations (UN) Secretary-General's Special Representative for Business and Human Rights (2005–11). His appointment to the UN position was a direct response to the unprecedented growth of the extractive industry globally and the concomitant escalation of violence and human rights abuses which accompanied it. Ruggie's proposal for addressing the situation are contained in the two renowned documents he authored: the "Framework" (2008) and the "Guiding Principles" (2011). Foundational to them, in terms of the conduct of business vis-à-vis human rights protections, is the principle of voluntarism, the gospel of the private sector when it comes to regulation of its conduct. As Ruggie wrote, "the responsibility of business [is] to respect human rights," (Ruggie 2011, 13). while states have a "duty" to protect such rights (Ruggie 2011, 3). Key here is that Ruggie's prescription in no way curbs or restricts the power of the private sector. Not surprisingly, therefore, his recommendations have become the new mantra for businesses seeking to project a responsible image (Deonandan and Morgan 2016) and many have comfortably declared their adherence to the principles he laid out. Detractors, however, have denounced these recommendations as setting the standards for human rights protections very low (Blitt 2012).

Ruggie's appointment to the Advisory Board, nevertheless, confers a significant degree of clout and prestige on the body and greatly enhances the national and global image of the Liberal government as a guarantor of human rights. It also simultaneously reassures the private sector that they are not in danger of being reigned in, restricted, or disciplined. As Moore (2018) cynically commented, "he [Ruggie] privileged the idea of private remedy that would be controlled by the company ... [and] we've seen how that's ... worked in Porgera[4] ... [I]f we think that really serious changes are going to come ... we're kidding ourselves." Moore's comments foreshadowed subsequent developments around CORE.

Another area of contention between CSOs and the government with respect to CORE stemmed from the latter's procrastination in appointing an Ombudsperson and in delineating CORE's powers. Civil society allies, such as KAIROS, which had given the administration plaudits for its extensive consultation processes, made public their dissatisfaction with this situation and launched a lobbying campaign demanding action on the issue (see KAIROS 2019a and 2019b). Their efforts were met with silence, that is, until April 2019, the final year of Liberal's first term. Then, the government announced its selection of Sheri Meyerhoffer a lawyer who had previously worked as a consultant and lobbyist in the oil and gas sectors (Government of Canada 2019b) and who now, as Ombudsperson, was tasked with oversight of the same sectors (McGee 2019).

While it acted on the appointment front, albeit after much prodding from civil society, the Liberal government is still shelving the decision on CORE's powers. The procrastination drove the entire contingency of seven civil society representatives and their alternates (a total of fourteen individuals) to resign from the advisory board three months after Meyerhoffer was installed, leaving only Ruggie along with industry and government representatives. According to one of the organizations, the CNCA (quoted in Lim 2019), this action was a result of the "the erosion of ... trust and confidence in the government's commitment to international corporate accountability," specifically its lack of empowerment of CORE. The organizations identified several problems that undermine the agency's oversight capacity.

To begin, they charged that CORE is lacking independence from the government as the latter has declined to use the Public Inquiries Act to render the organization free of official influence. Instead, the Ombudsperson and her staff are public servants within Global Affairs Canada reporting to the Minister of Small Business, Export Promotion and International Trade (CNCA 2020).

In addition, the groups have expressed concern that CORE may be susceptible to corporate influence as, "it appears that complainants may be required to use a process called joint fact-finding [that is they must collaborate with corporations in the process]. This would give already powerful companies an effective veto over what issues are looked at, who examines them, and whether any information is made public" (CNCA 2020). Very importantly too, and in keeping with corporate preferences, CORE is devoid of any investigative authority, enforcement powers, or any other form of remedies where abuses are alleged (Schreiber 2019). This too privileges corporations. Instead, CORE is empowered only to mediate and conduct "reviews," and these functions "rely entirely on the good will of transnational corporations to voluntarily share vital information that might implicate them in wrongdoing" (CNCA 2020). MiningWatch's Catherine Coumans attributed CORE's toothlessness directly to corporate lobbying, declaring that, "It is pretty clear what happened if you look at the lobby registry. They've [the government] been lobbied to death." (quoted in Friedman 2019). Concurring, Karyn Keenan of the non-profit Above Ground observed: "I think industry is profoundly uncomfortable with this office [CORE] and I think industry has made that known to government and I think that government has weakened the office in response to pressure from the industry" (quoted in Blanchfield 2019). Indeed, according to Friedman (2019), available data reveal that the industry organization, the Mining Association of Canada (MAC) "lobbied [the government] extensively on 'international trade' issues." MAC's president, Ben Chalmers, seemed to confirm these claims when he declared that his association is not in favour of an empowered CORE but prefers the body to focus on dispute resolution (Friedman 2019). His comments echo those of the new Ombudsperson herself, who has expressed a preference for the "collaborative approach" with businesses (quoted in Friedman 2019). This stance is no doubt comforting to the private sector. Reassuring to them also would be Meyerhoffer's background and her links to the energy industry. According to her LinkedIn profile, she did a master's degree in public administration at Harvard, where she was a student of John Ruggie, the Head of the Advisory Board and, more famously, the architect of voluntary governance schemes for the private sector. In addition, she worked in different capacities, as a consultant, lobbyist, and legal counsel to the energy sector (Meyerhoffer n.d.).

As expected, the makeup and mandate of CORE has led to widespread dissatisfaction among CSOs that had eagerly anticipated a decisive policy turn from Harper's insubstantial CSR Counsellor position. MiningWatch's Catherine Coumans confessed that she had believed the Liberals' promise about creating an office that "would possess authority to compel testimony or subpoena documents," all powers she deemed "critical" if accusations of corporate misconduct overseas are to be effectively investigated (quoted in Friedman 2019). KAIROS (2019c) voiced the sentiments of many in stating that it was "profoundly disappointed" by the Liberals' reneging on their earlier promise "to create an independent office with real powers to investigate abuses and redress harms caused by Canadian companies operating abroad." The group dismissed the Ombudsperson's role as "a powerless advisory post, little different from the Corporate Social Responsibility (CSR) Counsellor position that previously existed and [that] was abolished because it was deemed ineffective" (KAIROS 2019c). The CNCA (2020) was even stronger in its condemnation and basically dissuaded communities from resorting to it: "Given the blatant about-face of the … [Liberal] government on its commitments regarding the establishment of the CORE, there is currently no reason for communities to believe it will prove to be more effective than the offices that have preceded it."

While the discontent by CSOs with the Liberals on this issue does not reflect well on the government's commitment to abiding by its promises, it also placed the former in a political predicament. Given their acrimonious relationship with the Conservatives under Harper, their political options were limited in the 2019 elections as the long-term survival of many of them was strongly dependent on their ongoing receipt of governmental financial support. Displeasure with the Trudeau government did not automatically translate into their withdrawal of support from it. Siding with the Conservatives was not an option given the ideological orientation of the party and its earlier attempts to neutralize the sector. Further, the other alternative, the New Democratic Party (NDP), had almost no chance of forming the next government, according to the polling data prior to the 2019 election (Nanos 2019). As such, progressive civil society groups had only limited political manoeuvrability; they could criticize the Liberals, but in fact, the party was their only viable political option in practical terms. Indicative of this was the comment by Dwyer of the CNCA. She confessed that her organization would be willing to return to the Advisory Board if the Liberals fulfilled the promises they made to empower CORE (Lim 2019). Organizations such as KAIROS and CNCA continue to criticize the government on the matter and are actively campaigning for changes to CORE, but they seem to be doing so in an environment where they do not feel under threat of being punished.

For its part, despite the criticisms it has received over CORE's anemic powers, the government still has not altered course in terms of empowering the office, and it is unlikely that it will do so soon. Moore (2018) concurs with this in her assessment of CORE:

> We know that at the end of the day the recommendations to government and to companies out of that Ombudsman's office will still be non-binding … I don't believe [the Ombudsperson] is going to solve the problem by any means. I think that's why it was a fairly safe step for the government to take. Because at the end of the day it doesn't go to the core of the problem in terms of addressing impunity and in terms of addressing this system that privileges corporate interests over everything else. The fact that the Trudeau government is still looking for every opportunity to strike new free trade and investor protection agreements throughout Latin America with Mercosur, etcetera, still promoting Canadian corporate interests in some of the worst and most deadly parts of the region, I think is a signal that very little has really shifted … I think the government is caught in a bunch of contradictions, on a number of … [fronts].

These contradictions have no doubt been exacerbated by the COVID-19 pandemic. The crisis has wrought devastating economic consequences – a shrinking economy, extensive business closures, drastic declines in consumer spending, massive job losses, plunging domestic production, and steep declines in exports. Conversely, accompanying the economic contraction, is the critical need for increased governmental spending to keep vulnerable citizens and businesses afloat in the crisis. This dismal economic picture has pushed discussions on CORE's power defects to the margins for the government; its focus in the wake of the COVID-19 crisis is on rebuilding the private sector and not restraining it.

Funding

Contradictions and back-tracking are also reflected in the Liberal government's funding for overseas development assistance. Soon after securing his first and decisive electoral victory, Trudeau declared to Canadians and the world that a "more compassionate" Canada "is back" (quoted in Gullies 2015). For those NGOs involved in overseas development work, and particularly those defunded by the Harper government, this was undoubtedly a reassuring signal.

The Liberals soon followed through on the commitment implied in the Prime Minister's statement and set in motion a process of inclusivity and engagement with many of the groups that had been targeted by the Conservatives. As mentioned earlier, there were extensive processes of consultation to solicit their input on policy, and they were all eligible for funding. What did not change, however, was the funding format which continued Harper-era policies.

Rather than have ongoing guaranteed funding at a particular level, organizations now had to compete for existing moneys through the government's call for proposals for projects or were invited to apply by means of "shoulder-tapping." As KAIROS's Jennifer Henry (2018c) related: "What has changed in Global Affairs (from CIDA) is that there appears to be no such thing as ongoing organizational program funding. When we were defunded it was really [of] an expected renewal (we had been funded in one iteration or another for 34 years). What the Harper government brought in was funding only through response to 'call for proposals [CFP]' ... But what began to open up [under the Liberals], more publicly, was the possibility for 'unsolicited proposals' and that is what we applied under ... Now there is a combination of unsolicited, shoulder tapping and call ... [for] proposals."

This strategy has many drawbacks for the NGO sector. They are put in competition with each other for a finite amount of dollars, and funding is likely to go to that group with the most significant expertise in the area of the CFP. This could mean that the largest groups emerge successful more frequently, as they are likely to have a greater array of expertise.

Further, this process of responding to CFP means that development organizations are constantly writing and rewriting applications, which can be extremely time consuming and which utilizes valuable resources that are in short supply for many of the groups. This bureaucratic distraction has echoes of the toll the CRA audits exacted on the organizations in terms of the time and resources they absorbed. Referencing KAIROS's recent success for a project to empower grassroots women's groups (discussed in greater detail below), Henry (2018c) revealed that that proposal took over two years as their application had to be reworked and resubmitted before funding was approved. In addition, what is unknown is what the prospects are for groups whose activities do not fall within the ambit of the government's area of preferences (Brown 2018b). Given that the lion's share of available dollars will likely go to organizations whose priorities coincide with that of the government, this strategy is an implicit mechanism of control. It has the potential to either restrict the activities of groups in the sector or compel them to redirect their focus to areas deemed more appropriate by the state.

Aside from the control it exerts, this approach to funding also does not address the issue of insecurity and instability of NGO partners in the global South. The CCIC had recommended that funding rules be changed so that a genuine partnership between Canadian NGOs and those in the global South can be created. In other words, the rules need to allow for a long-term development focus rather than short term. The funding policy instituted by the Liberals does the converse. Since financial support is now based on short-term contracts, the organizations are compelled to devote significant resources to the application process while simultaneously existing in a state of perpetual financial insecurity. Henry (2018c) explained the precarity thus: "This is a project which will end after 5 years unless we apply for and negotiate an overlapping project."

Where Harper implemented funding cuts and CRA audits to circumscribe the activities of civil society groups, the Trudeau government's strategy, while different, nevertheless, still results in financial insecurity, and the groups that are likely to be successful are those with missions and goals that coincide with those of the government. It can be argued that what Harper's Conservatives accomplished with sticks, Trudeau's Liberals secured with carrots. The "carrot" technique was also evident in the latter's promotion of their gender equality, as elaborated on below.

Gender Equality and the Feminist International
Assistance Policy (FIAP)

One of the Liberal government's policy innovations that won praise from progressive civil society, including those which had been targeted and ostracized under Harper (such as MATCH and KAIROS), was the government's launch of its ambitious "Feminist International Assistance Policy (FIAP)"in June 2017. Proponents of the policy were particularly pleased not only by what they saw as progressivism in the language of the policy but also in the funding provided to support the policy's objectives. Detractors such as Alexandra Dobrowolsky (2020), however, interpreted Trudeau's feminist agenda as a form of rollout neoliberalism: "The feminism that Trudeau would go on to promote in office—one that set goals around prosperity and leadership is activated in policies weighted towards economic returns—would therefore only improve the status of a particular stratum of already privileged women. Despite rhetorical comments to equality and diversity, we see real gaps in intersectionality in the promotion of a feminism that is more in keeping with neoliberal feminist principles. A focus on prosperity is based on a classical liberal assumption that the economic realm is a place of opportunity. That is not the reality for most women" (31).

Drafted after extensive consultations with a wide range of CSOs (over 1,500 reported participated) (Brown and Swiss 2018), FIAP's objective is to promote "gender equality and the empowerment of women and girls" (GAC 2017), and it is considered among the first of its kind globally. This legislation represents a marked turn from Harper-era polices that had purposefully eschewed progressivism on gender issues. As Tiessen and Carrier (2015) reminded us, the Conservatives had even sought to alter the terminology in official documents to downplay and even eliminate references to "gender equality," ordering that the term be eradicated and replaced with "equality between women and men."[5] In explaining civil society's very positive response to FIAP, Beth Woroniuk of the Equality Fund (formerly MATCH International Women's Fund) stated, "We saw this as a marked departure from previous approaches to women's rights and gender equality where [the latter went] ... from being a cross-cutting theme to actually the major focus. So we saw a change in the language and the direction of the policy. We saw actually putting feminist in the title as a significant move especially globally when women's rights are being threatened and ... civil society's space is being shrunk in many cases. When there are more threats against women human rights defenders, it's a bit of a promising direction" (Woroniuk 2018).

FIAP has also won praise from CSOs for the funding backing the policy. The Liberals pledge that by 2021–2, "95% of bilateral international development assistance initiatives will target or integrate gender equality and the empowerment of women and girls (Government of Canada 2019c), a level of financing that will far exceed that of other [Organisation for Economic Cooperation Development Assistance Committee (OECD/DACDAC)] donors' gender programs" (Brown and Swiss 2018, 120).

Additionally, there were several different categories or sources of funding for FIAP-related initiatives. One of these is the Women's Voice and Leadership (WVL) programme to support local organizations that advance women's rights in developing countries and to which the government committed $174 million (WRPG/CCIC 2020), increased from the initial announcement of $150 million. The programme is currently operational in over thirty countries, including those in Latin America (WRPG/CCIC 2020). Praise for the policy came from many groups, including from those organizations that had been sidelined by Harper. Woroniuk, for example, pointed to the policy's direct financial support for grassroots organizations in the developing world as a major step:

> [T]here are specific mentions in the policy of the importance of women's rights organizations, which is really, really important because one of the things that we distinguish is funding, general funding for gender equality programming which could be done by the UN or done by a large NGO … [T]those programs are valuable but they are different than funding that goes directly to feminist organizations, women led organizations, that contribute to movement building.
>
> We see that as really, really key. That's a fundamental part of the kind of work that we do and what we believe in. And this policy does mention the importance of that kind of work so the door [is] opened in the policy to increase the level of direct funding going to women's rights organizations.
>
> There was, in the policy the announcement of a $150 million initiative. aimed at women's rights organizations … [T]hat was … huge. We've never seen $150 million for women's rights organizations before so that was really important.
>
> (Woroniuk 2018)

Glowing endorsement also came from the Nobel's Women's Initiative (NWI), which lauded the government both for the policy mission and the funding commitment. In a joint press release, NWI and MATCH thanked the government and commended it for what they hailed as a "bold" initiative: "Today, the Canadian government announced $150 million in funding for grassroots women making bold change—the largest single boost of funding from a government for women's rights! Thank you, Justin Trudeau and Minister Bibeau for championing feminist international development and showing the world that Canada truly is a global leader" (MATCH 2018).

While challenges have been identified with WVL, including that larger organizations and those based in Canada are privileged by the selection process due to their access to resources to complete the onerous application requirements, among other factors (WRPG/CCIC 2020), the programme has strong support among CSOs (see WRPG/CCIC 2020), as the aforementioned comments illustrate.

The Trudeau Liberals garnered kudos from civil society for other initiatives related to the advancement of women as well. KAIROS, another of the groups caught in Harper's disciplinary campaign, was revitalized under the Liberals (this despite the group's ongoing lobbying efforts to get the government to empower CORE). The group received a commitment from the government of $4.5 million over five years to help grassroots women's organizations in five countries (the Philippines, South Sudan, Colombia, and the West Bank), under its "Women of Courage: Women, Peace and Security" programme (KAIROS 2018).

The centrepiece of FIAP, however, and that which strongly emphasizes the role of the private sector in fostering women's empowerment, is the Equality Fund established in June 2019. This programme was conceived of through a combined process of extensive stakeholder consultation and in response to a call for proposal (CFP) on the "partnership for gender equality." The Fund consists of a consortium of civil society groups (both Canadian and international), community foundations, financial institutions, venture capitalists, philanthropists, and governments. The strong private-sector presence was an important requirement in the CFP (see Government of Canada 2019f), and the winning bid was submitted by then MATCH International Women's Fund, another of the organizations Harper had targeted and which almost folded as a result. Having won the bid, MATCH subsequently changed its name to the Equality Fund.

The Fund's goal, according to Global Affairs (GAC 2019) is to enable member organizations to "pool and leverage resources to attract even more new partners and create opportunities to close gender gaps and eliminate barriers to gender equality."

What is unique about the Equality Fund is that it was allocated the historic sum of $300 million[6] (Government of Canada 2020) and is thus considered "world's largest women's fund" (The Equality Fund 2020a). The Fund's web page describes it as "groundbreaking" and declares it "the single largest investment that one country has ever made to women around the globe" (The Equality Fund 2020b). By the end of 2019, the Fund had already reached $100 million and had a target of $1 billion over the next fifteen years (GAC 2019). Civil society partners greeted the government's actions with superlatives. Jess Tomlin (quoted in Chung 2020), formerly president and CEO of MATCH, now the fund's co-CEO, pronounced it "one of the most exciting initiatives ... because ... [it is] going to solve ... some of the most significant issues of our time." Tomlin's co-CEO, Jessica Houssain of Women Moving Millions, proclaimed that "This is an historic moment in the fight for equality around the world" (quoted in GAC 2020). For these civil society groups, what is historic too about the fund is not just the size of its dollar allocation but the predictability of it, as well as its promise of support directly to women's groups in developing countries. Tomlin and Houssain explained that the Fund has altered the prior "charitable model" approach to gender empowerment, one that emphasized "service delivery," was "project-based and piecemeal," and did not "allow local organizations to determine their own priorities."

Although the Equality Fund has met with widespread praise from supporters, an effective assessment is not feasible given that it is in its nascent stages at the time of writing. However, what should be noted is that while this programme is indeed innovative on many fronts, and while it may merit some of the superlative descriptors which greeted it, it is an approach to gender empowerment that is very much in keeping with a market-oriented development strategy. For example, as McGill and Munro (2020) observed, it is a business model of funding which stresses "legacy investments" to ensure that support for gender empowerment is not reliant on government financing. This raises concerns regarding the independence of the Fund from investors, the nature of investments that will be pursued, and how the resources will be distributed, among other issues (McGill and Munro 2020). Importantly, actions necessary for true gender empowerment are not of a short-term nature, yet investors prefer to see profits on investments in the short term. This led McGill and Munro (2920) to ask: "How will tough decisions be made regarding [types of investments and] investment returns, especially in the context of enormous financial risk and volatility, which may not be short term in nature?" Relatedly, the Fund incorporates language such as "best practices" (McGill and Munro 2020), buzzwords, or terms associated with neoliberal criteria for organizational evaluation.

Conversely, as with the apprehension over the degree of independence from investors, there are also similar concerns over the level of independence from government. Given that the Canadian government is the primary investor thus far, providing approximately 75 per cent of the contributions, this raises the question as to whether Fund partners will be free to decide on policy directions without governmental interference (McGill and Munro 2020). Hence, while the government may speak of empowering local women's organizations, in fact, the Equality Fund carries within it a strong indication of a top-down approach, and inherent in it is the likelihood that it will be driven by the interests of its investors.

More general criticisms relate to the rhetoric around the Fund. Observers have labelled some of the tributes as "hyperbolic" and "grandiose" in light of the fact that the Fund anticipates issuing "only $80 million in grants in its first four years while taking 15 years to raise an endowment of $1 billion" (McGill and Munro 2020). Nevertheless, the Equality Fund represents a major advance from Harper-era policies, especially when it comes to gender empowerment both at home and abroad. Despite their critical analysis of the fund, development scholars McGill and Munro (2020) suggest it is "an experiment worth watching". However, one should not lose sight of the ideology guiding the funds composition and modus operandi.

Criticisms of the Liberals' market approach to gender empowerment also relates to their creation of the Development Finance Institute Canada (FinDev), which was established in 2017 and assigned a budget of $300 million. While the agency is separate from FIAP, its mandate is related to it. FinDev's mission is "to provide easier access to capital for entrepreneurs in developing markets," namely, in Latin America, the Caribbean, and sub-Saharan Africa, to address issues of poverty reduction, climate change, and job creation, and to promote women's economic empowerment (FinDev n.d.a). As the organization's website states: "The private sector can play a significant role in increasing economic equality between men and women and spurring economic empowerment" (FinDev n.d.b).

FinDev was initially conceived of under the Harper administration, and both leaders gave similar rationale for its creation. They argued that since fiscal challenges demand that overseas development assistance (ODA) levels be contained, the current funding has to deliver more effective results, and this can only be done through the private sector. FinDev's mission then is to "grow a financially sustainable portfolio … focusing on the needs of private sector borrowers" and it promises to do so while demonstrating the "viability of gender-lens investing" (FinDev n.d.b). In other words, FinDev's strategy is to rely on the private sector and market forces for women's empowerment in such areas as "women's entrepreneurship … [their] access to leadership positions, … [and]

to quality employment [and] basic services" (FinDev n.d.b). However, given that profit maximization and promoting shareholder interests are goals of the private sector, it remains unclear how the Institute will promote women's empowerment in this context (Brown and Swiss 2018). As Brown and Swiss (2018) observed with regard to the limits on women's empowerment that could result due to private-sector involvement in the programme: "Corporate social responsibility and other voluntary charitable projects may generate some benefits for marginalized and disadvantaged people, but those are side activities, not core ones.... [I] t will be a major challenge for the Canadian government to ensure that the benefits from its promotion of the private sector accrue primarily to women and reduce gender inequalities. It will also be harder to ensure accountability, especially in cases of 'blended finance'. The use of loans, rather than grants, could leave beneficiaries worse off if their ventures fail and they must still repay the capital provided from the Canadian government, with interest" (126).

Conclusion

Prime Minister (PM) Justin Trudeau won plaudits on the global stage for his many policy innovations after his electoral victory in his first national political campaign. These accolades extended to his initiatives with respect to civil society, which included his gender-balanced cabinet, his establishment of the Ombudsperson for Responsible Enterprise, and his adoption of the Feminist International Assistance Policy, among others. He even inspired the French President Emmanuel Macron who modelled his own gender-balanced cabinet after that of the Canadian PM. Trudeau also has garnered global recognition for his progressive rhetoric on the very pressing issue of reconciliation with Indigenous peoples. Seemingly, these all stand in stark contrast to the policies and attitudes of his predecessor Stephen Harper whose blunt, unapologetic, pro-corporate agenda alienated and marginalized many civil society groups, especially the progressive organizations.

However, the foregoing analysis of some of Trudeau's policy stances and initiatives vis-à-vis civil society groups, indicate that although he undoubtedly represented a departure from the Harper era (for example, in terms of consultation with and funding for this sector), he nevertheless largely maintained an agenda that was consistent on many fronts with that of the Conservatives. Jen Moore (2018) of MiningWatch argued that the notion of Trudeau as a transformative leader (in comparison to Harper) is unfounded, and she cites Canada's ongoing policies on behalf of Canadian mining investments in Latin America and else-where and its pursuit of free trade policies as counterpoints. Other civil society groups were also becoming fearful during Trudeau's first man-date that a renewed advocacy chill was on the horizon as the govern-ment had failed to halt many of Harper's anti-NGO tactics, delayed appointing an Ombudsperson for CORE, and when the appointment was made there were no efforts to empower the office.

Perhaps it was Liberal awareness of the simmering dissatisfaction among civil society groups coupled with their declining poll numbers, as the second election loomed in 2019, that led the government to mod-ify its stance on several fronts. In quick succession, it adopted a series of measures demanded by civil society, including cancelling the CRA audits, amending the Income Tax Act (to remove references to political activities of non-profits), establishing the office of CORE, and appoint-ing an Ombudsperson.

While not all the actions satisfied the progressive civil society sector, the latter's electoral options were limited. The Conservative platform did not reflect their preferences, and the New Democratic Party, while likely in alignment with their agenda, was a distant third in the polls (Nanos 2019) and unlikely to win power. At the same time, progressiv-ism was under threat as Conservative governments were elected across the country in provincial contests in British Columbia, Alberta, Saskatch-ewan, Manitoba, Ontario, and New Brunswick. Not surprisingly, there-fore, despite their disenchantment with the Liberals on several fronts (its pursuit of free trade and its weak response to empowering CORE, for example), civil society organizations did not come out against the Liberals during the campaign. In the end, while Trudeau did not win decisively, he nevertheless was victorious, albeit with a minority gov-ernment. What this might imply for CSOs is as yet unclear, though the recent funding announcements and the creation of the Equality Fund would indicate that he is generally supportive of their agenda.

However, what has remained consistent, is that both in their first and second terms, the Liberals have articulated policy stances vis-à-vis civil society groups that combined market-type initiatives with progressive rhetoric, and in this way they have differentiated themselves from their predecessor. Commenting on Trudeau's particular style (in the wake of his first electoral victory and the global accolades he was receiving), and referencing the earlier global popularity of the current PM's father during the 1960s–1980s and the phenomenon that bore his name, Por-ras Ferreira (2016) wrote in the *New York Times*, "Trudeaumania moves forward with specific steps, a fresh style and velvety words."

NOTES

1 The authors wish to thank Paul Haslam and the editors of this volume
   for their constructive comments on an earlier draft of the article. This
   research was supported by a grant from the Social Sciences and Humanities
   Research Council of Canada (SSHRC).
2 The terms civil society organizations (CSOs), non-governmental
   organizations (NGOs), non-profits, and charities are used interchangeably
   in this chapter.
3 In 2016, Trudeau received a letter from over 180 civil society organizations
   calling for better regulation of Canadian mining firms (AIDA 2016).
4 This is a reference to the violence and environmental destruction at Barrick
   Gold's Porgera mine in Papua New Guinea.
5 Some scholars, however, have expressed concern over FIAP's terminology.
   Mason (2020, 4–5) has criticized the broad use of the concept of
   "intersectionality," pointing out that not only is it "not clearly defined"
   but it is utilized to imply "inclusion" and ignores the powerful and
   intertwining forces that intersectionality addresses. For Mason (2020, 14),
   FIAP's reductive usage of the term suggests the government's invoking of
   a buzzword, a strategy that runs the risk of "instrumentaliz[ing] ... feminist
   concerns toward other development objectives."
6 Another funding dimension to FIAP is the government's announcement
   that it will provide up to $30 million in matching federal funds to
   organizations working for women's rights in Canada (see Government of
   Canada 2019g).

REFERENCES

AIDA (Asociación Interamericana para la Defensa del Ambiente/
   Interamerican Association for Environmental Defense). 2016. "Open Letter
   on Mining to Canadian Prime Minister Trudeau." Accessed 27 October
   2021. https://aida-americas.org/sites/default/files/publication/letter_to
   _trudeaueng.pdf.
Amnesty International. 2017. "Defending Rights for All? A Human Rights
   Assessment of Year One of the Trudeau Government." Accessed 27 October
   2021. https://www.yumpu.com/en/document/read/56528742
   /defending-rights-for-all-international.
Ayers, Allison. 2006. "Demystifying Democratisation: The Global Constitution
   of (Neo)liberal Politics in Africa." *Third World Quarterly* 27, no. 2: 321–38.
Barrick. 2018. "Barrick Reports 2017 Full Year and Fourth Quarter Results."
   Press release (14 February). Accessed 27 October 2021. https://www

.barrick.com/news/news-details/2018/Barrick-Reports-2017-Full-Year-and
-Fourth-Quarter-Results/default.aspx.

Barry-Shaw, Nikolas, and Dru Oja Jay. 2012. "NGOs and Empire." *Briarpatch*
41, no. 4: 31–4.

Beeby, Dean. 2014. "Canada Revenue Agency Accused of 'Political' Targeting
of Charities." *Macleans*. Accessed 25 November 2021. https://www
.macleans.ca/news/canada/canada-revenue-agencys-political-targeting
-of-charities-under-scrutiny/.

– 2016a. "Charities Push Back against Liberals on Political Audits." CBC
News (14 March). Accessed 27 October 2021. https://www.cbc.ca/news
/politics/charities-push-back-against-liberals-on-political-audits-1.3490988.

– 2016b. "Anti-poverty Group Launches Challenge of Political-Activity
Limits." CBC News (6 September). Accessed 27 October 2021. https://
www.cbc.ca/news/politics/charities-political-activities-canada-revenue
-agency-canada-without-poverty-charter-legal-1.3744919.

– 2017. "Political Activity Audits of Charities Suspended by Liberals." CBC
News (4 May). Accessed 27 October 2021. https://www.cbc.ca/news
/politics/canada-revenue-agency-political-activity-diane-lebouthillier
-audits-panel-report-suspension-1.4099184.

– 2018a. "Charities 'Worried' after Meeting with Morneau on 'Political
Activity' Law." CBC News (19 April). Accessed 27 October 2021. https://
www.cbc.ca/news/politics/liberal-charity-income-tax-1.4624600.

– 2018b. "Liberals Promise to Repeal Restriction on Charities' Political
Activity." CBC News (15 August). Accessed 27 October 2021. https://www
.cbc.ca/news/politics/cra-court-charities-political-activity-1.4786818.

– 2019. "Ottawa Drops Appeal in Political Activity Case, Ending Charities'
7-year Audit Nightmare." CBC News (31 January). Accessed 27 October
2021. https://www.cbc.ca/news/politics/cra-revenue-charities-freedom
-expression-lebouthillier-appeal-political-activities-court-1.5001087.

Blanchfield, Mike. 2019. "Rules for New Corporate Ombud Criticized for
Deferring to Companies." *Canadian Press* (11 June). Accessed 27 October
2021. https://www.nationalobserver.com/2019/06/11/news/rules-new
-corporate-ombud-criticized-deferring-companies.

Blitt, R.C. 2012. "Beyond Ruggie's Guiding Principles on Business and Human
Rights: Charting an Embracive Approach to Corporate Human Rights
Compliance." *Texas International Law Journal* 48, no. 1: 33–62.

Bone, Jeffrey. 2015. "The State of Canada's Corporate Social Responsibility
Strategy." OpenCanada Org. (25 March). Accessed 27 October 2021.
https://www.opencanada.org/features/the-state-of-canadas-corporate
-social-responsibility-strategy/.

Brenner, N., J. Peck, and N. Theodore. (2010). "After Neoliberalization?"
Globalizations 7, no. 3: 327–45.

Bridge, Richard. N.d. "Making Sense of the New 'Canada Not-for-Profit Corporations Act.'" Accessed 27 October 2021. http://rcafassociation.ca /uploads/airforce/2012/03/Making-Sense-of-the-New-Canada-Not-for -profit-Corporations-Act-Hightlights-November-Session.pdf.

Brown, Stephen. 2016. "Undermining Foreign Aid: The Extractive Sector and the Recommercialization of Canadian Development Assistance." In *Rethinking Canadian Aid*, 2nd ed. edited by Stephen Brown, Molly den Heyer, and David Black, 277–96. Ottawa: University of Ottawa Press. http://aix1.uottawa.ca/~brown/pages/Stephen%20Brown%20 Undermining%20Foreign%20Aid.pdf.

– 2017. "Harper Lite? The Trudeau Government on Foreign Aid." *Centre for International Policy Studies* (2 April 2). Accessed 27 October 2021. http:// www.cips-cepi.ca/2017/04/02/harper-lite-the-trudeau-government-on -foreign-aid/.

– 2018. "All about That Base? Branding and the Domestic Politics of Canadian Foreign Aid." *Canadian Foreign Policy Journal* 24, no. 2: 145–64.

Brown, Stephen, and Liam Swiss. 2018. "Canada's Feminist International Assistance Policy: Bold Statement or Feminist Fig Leaf?" In *How Ottawa Spends, Canada@150, 2017–2018*, edited by Katherine A.H. Graham and Allan M. Maslove, 117–31. Ottawa: Carleton University. Accessed 27 October 2021. https://carleton.ca/hos/wp-content/uploads/How -Ottawa-Spends-2017-2018-Ottawa-@150.pdf.

Chung, Lance. 2019. "The Equality Fund is Redefining Philanthropy by Empowering Women." *Bay Street Bull*. Accessed 27 October 2021. https:// www.baystbull.com/the-equality-fund-is-redefining-philanthropy-by -empowering-women/.

CNCA (Canadian Network on Corporate Accountability). 2016. "Letter to the Honourable Stéphane Dion." 4 January. Accessed 27 October 2021. http:// cnca-rcrce.ca/wp-content/uploads/2016/05/Correspondence-between-the CNCA-and-Minister-Dion-January-to-May-20161.pdf.

– 2020. "Canadian Ombudsperson for Responsible Enterprise (CORE): Approach with Caution – Canadian Civil Society Groups Raise Alarm about Canada's Ombudsperson for Responsible Enterprise (CORE)." 30 April. Accessed 27 October 2021. http://cnca-rcrce.ca/wp-content /uploads/2020/04/core-caution-E-1.pdf.

Conflictos Mineros en América Latina – Observatory of Mining Conflicts in Latin America (OCMAL). 2020. Accessed 28 October 2021. https://mapa .conflictosmineros.net/ocmal_db-v2/conflicto.

"CRA Charity Audits." n.d. Accessed 27 October 2021. http://cponline .thecanadianpress.com/graphics/2014/cra-audit-charities/index.html.

Deonandan, Kalowatie, and Jennifer Morgan. 2016. "The Privatization of Human Rights: The Human Rights Impact Assessment at Goldcorp's Marlin Mine in Guatemala." In *Mining in Latin America: Critical Reflections on the New Extraction*, edited by Kalowatie Deonandan and Michael Dougherty, 160–82. London and New York: Routledge.

Dobrowolsky, Alexandra. 2020. "A Diverse, Feminist 'Open Door' Canada? Trudeau-Styled Equality, Liberalism and Feminism." In *Turbulent Times, Transformational Possibilities: Gender and Politics Today and Tomorrow*, edited by Fiona MacDonald and Alexandra Dobrowolsky, 23–48. Toronto: University of Toronto Press.

The Equality Fund. 2020a. "Equality Fund: Funding Feminist Futures." Accessed 28 October 2021. https://equalityfund.ca/.

– 2020b. "Our Herstory." Accessed 28 October 2021. https://equalityfund.ca/who-we-are/our-herstory/.

FinDev. n.d.a. "Our Mission is to Make an Impact." Accessed 27 October 2021. https://www.findevcanada.ca/en/who-we-are/our-story.

– n.d.b. "Our Gender Lens Approach." Accessed 27 October 2021. https://www.findevcanada.ca/en/what-we-do/gender-lens-investing.

Friedman, Gabriel. 2019. "'Lobbied to Death': Liberals Face Backlash over Corporate Responsibility Ombudsman." 8 April. Accessed 27 October 2021. https://business.financialpost.com/commodities/mining/lobbied-to-death-liberals-face-backlash-over-corporate-responsibility-ombudsman.

GAC (Global Affairs Canada). 2017. "Canada's Feminist International Assistance Policy." Accessed 27 October 2021. https://international.gc.ca/world-monde/assets/pdfs/iap2-eng.pdf.

– 2019. "Global Affairs Canada - The Equality Fund: Transforming the Way We Support Women's Organizations and Movements Working to Advance Women's Rights and Gender Equality." Accessed 27 October 2021. https://www.canada.ca/en/global-affairs/news/2019/06/global-affairs-canada---the-equality-fund-transforming-the-way-we-support-womens-organizations-and-movements-working-to-advance-womens-rights-and-g.html.

– n.d. "Responsible Business Conduct Abroad – Questions and Answers: Canadian Ombudsperson for Responsible Enterprise (Ombudsperson)." Accessed 27 October 2021. http://www.international.gc.ca/trade-agreements-accords-commerciaux/topics-domaines/other-autre/faq.aspx?lang=eng.

Goar, Carol. 2014. "Stephen Harper Intimidates Charities into Silence: Goar." Commentary. *The Star* (15 July). Accessed 27 October 2021. https://www.thestar.com/opinion/commentary/2014/07/15/stephen_harper_intimidates_charities_into_silence_goar.html.

Gordon, Todd. 2012. "Canadian Development Aid Takes on Corporate Colouring." Editorial Opinion. *Toronto Star*. (29 November). Accessed 23 November 2021. https://www.thestar.com/opinion/editorialopinion /2012/11/29/canadian_development_aid_takes_on_corporate _colouring.html.

Government of Canada. 2007. "Prime Minister Harper Signals Canada's Renewed Engagement in the Americas." Accessed November 25, 2021. https://www.canada.ca/en/news/archive/2007/07/prime-minister -harper-signals-canada-renewed-engagement-americas.html.

– 2011a. "Minister Oda Announces Initiatives to Increase the Benefits of Natural Resource Management for People in Africa and South America." 29 September. Accessed 27 October 2021. https://www.canada.ca/en /news/archive/2011/09/minister-oda-announces-initiatives-increase -benefits-natural-resource-management-people-africa-south-america.html.

– 2011b. "Regulatory Impact Analysis Statement." Accessed 27 October 2021. http://gazette.gc.ca/rp-pr/p1/2011/2011-02-26/html/reg7-eng.html.

– 2016. "Minister Lebouthillier Announces Winding Down of the Political Activities Audit Program for Charities." 20 January. Accessed 27 October 2021. https://www.canada.ca/en/revenue-agency/news/2016/01/minister -lebouthillier-announces-winding-down-of-the-political-activities-audit -program-for-charities.html.

– 2017a. "Minister Lebouthillier Welcomes the Panel Report on the Public Consultations on Charities and Political Activities." 4 May. Accessed 27 October 2021. https://www.canada.ca/en/revenue-agency/news/2017 /05/minister_lebouthillierwelcomesthepanelreportonthepublic consultat.html.

– 2017b. "Consultations to Renew Canada's International Assistance." Accessed 27 October 2021. http://international.gc.ca/world-monde/issues _development-enjeux_developpement/priorities-priorites/consultation .aspx?lang=eng.

– 2017c. "Report of the Consultation Panel on the Political Activities of Charities." (31 March). Accessed 28 October 2021. https://www.canada .ca/en/revenue-agency/services/charities-giving/charities/about -charities-directorate/political-activities-consultation/consultation-panel -report-2016-2017.html.

– 2018. "Statement by the Minister of National Revenue and the Minister of Finance on the Government's Commitment to Clarifying the Rules Governing the Political Activities of Charities." (15 August). Accessed 27 October 2021. https://www.canada.ca/en/revenue-agency/news/2018 /08/statement-by-the-minister-of-national-revenue-and-minister-of -finance-on-the-governments-commitment-to-clarifying-the-rules -governing-the-political.html.

– 2019a. "The Government of Canada Delivers on its Commitment to Modernize the Rules Governing the Charitable Sector." Accessed 27 October 2021. https://www.canada.ca/en/revenue-agency/news/2019 /03/the-government-of-canada-delivers-on-its-commitment-to-modernize -the-rules-governing-the-charitable-sector.html.
– 2019b. "Minister Carr Announces Appointment of First Candian Ombudsperson for Responsible Enterprise." (8 April). Accessed 27 October 2021. https://www.canada.ca/en/global-affairs/news/2019/04/minister -carr-announces-appointment-of-first-canadian-ombudsperson-for -responsible-enterprise.html.
– 2019c. "Canada's Feminist International Assistance Policy." April. Accessed 27 October 2021. https://international.gc.ca/world-monde/issues _development-enjeux_developpement/priorities-priorites/policy-politique .aspx?lang=eng.
– 2019d. "Public Policy Dialogue and Development Activities by Charities: Guidance." Accessed 28 October 2021. https://www.canada.ca/en/revenue -agency/services/charities-giving/charities/policies-guidance/public -policy-dialogue-development-activities.html?utm_source=stkhldrs&utm _medium=eml&utm_campaign=PPDDA.
– 2019e. "The Government of Canada Announces Full Membership of the Advisory Committee on the Charitable Sector." Accessed 28 October 2021. https://www.canada.ca/en/revenue-agency/news/2019/08/the -government-of-canada-announces-full-membership-of-the-advisory -committee-on-the-charitable-sector.html.
– 2019f. "Request for Expressions of Interest to Design and Manage the Partnership for Gender Equality." Accessed 28 October 2021. https://www .international.gc.ca/world-monde/funding-financement/gender_equality -egalite_genres.aspx?lang=eng.
– 2019g. "Canada and Partners Announce New Legacy Investments to Support Women's Rights and Gender Equality at Home and Abroad." Accessed 28 October 2021. https://www.canada.ca/en/global-affairs /news/2019/06/canada-and-partners-announce-new-legacy-investments -to-support-womens-rights-and-gender-equality-at-home-and-abroad. html.
– 2020. "Partnership to Fund Gender Equality and the Empowerment of Women and Girls in Canada and Abroad." Accessed 28 October 2021. https://www.international.gc.ca/gac-amc/campaign-campagne/gender _equality-egalite_des_genres/index.aspx?lang=eng.
– n.d. "Who We Are." Accessed 28 October 2021. https://www.canada.ca /en/revenue-agency/services/charities-giving/charities/about-charities -directorate/who-we.html.

Harper, Stephen. 2007."Prime Minister Harper Signals Canada's Renewed Engagement in the Americas." Santiago, Chile (17 July). http://pm.gc.ca/eng/media.asp?id=1759.

Henry, Jennifer. 2018a. Executive Director, KAIROS: Canadian Ecumenical Justice Initiatives. Telephone interview by authors Kalowatie Deonandan and Toveli Schmuland, 26 April 2018.

– 2018b. Executive Director, KAIROS: Canadian Ecumenical Justice Initiatives. Email Communications with Author Kalowatie Deonandan, 23 August 2018.

ICAR (International Corporate Accountability Roundtable). 2018. "The Newest Corporate Watchdog: Canada Announces an Ombudsperson for Corporate Accountability." Accessed 28 October 2021. https://www.business-humanrights.org/en/latest-news/the-newest-corporate-watchdog-canada-announces-an-ombudsperson-for-corporate-accountability/.

Industry Canada. 2008. "Government of Canada Tables New Regime for Not-for-Profit Corporations." (3 December). https://www.ic.gc.ca/eic/site/cd-dgc.nsf/eng/cs05170.html.

KAIROS: Canadian Ecumenical Justice Initiatives. n.d. "Ecological Justice: Tar Sands." Accessed 28 October 2021. https://www.kairoscanada.org/what-we-do/ecological-justice/tar-sands.

– 2018. "Media Release: KAIROS' Women of Courage: Women, Peace and Security Program to receive $4.5 million." (30 May). Accessed 28 October 2021. https://www.kairoscanada.org/media-release-kairos-women-courage-women-peace-security-program-receive-4-5-million.

– 2019a. "Urgent Action: It's Time to Ramp Up the Pressure for an Ombudsperson!" (25 March). Accessed 28 October 2021. https://www.kairoscanada.org/urgent-action-time-ramp-pressure-ombudsperson.

– 2019b. "Time is Running Out for an Independent Ombudsperson with Power to Investigate…" (4 April). Accessed 28 October 2021. https://www.kairoscanada.org/letter-honourable-minister-carr-regarding-canadian-ombudsperson-responsible-enterprise-core.

– 2019c. "An Independent and Empowered CORE Is at the Core of Corporate Accountability and Gender Justice." (10 April). Accessed 28 October 2021. https://www.kairoscanada.org/independent-empowered-core-core-corporate-accountability-gender-justice.

Kingfisher, Catherine. 2002. "Introduction: The Global Feminization of Poverty." In *Western Welfare in Decline: Globalization and Women's Poverty*, edited by Catherine Kingfisher, 3–12. Philadelphia: University of Pennsylvania Press.

The Liberal Party of Canada. "Real Change: A New Plan for a Strong Middle Class." Accessed 28 October 2021. https://liberal.ca/wp-content/uploads/sites/292/2020/09/New-plan-for-a-strong-middle-class.pdf.

Lim, Jolson. 2019. "Civil Society, Labour Groups Resign in Protest from Federal Panel on Corporate Responsibility Abroad." (11 July). Accessed 28 October 2021. https://ipolitics.ca/2019/07/11/civil-society-labour -groups-resign-in-protest-from-federal-panel-on-corporate-responsibility -abroad/.

Mason, Corinne. 2019. "Buzzwords and Fuzzwords: Flattening Intersectionality in Canadian Aid." *Canadian Foreign Policy Journal* 25, no. 2:203–19.

MATCH International Women's Fund. 2018. "Social Media Toolkit." Press release. Accessed 28 October 2021. http://cdn.matchinternational.org /wp-content/uploads/2013/10/FeministFundAnnouncementSMKit -FINAL.pdf.

McGee, Niall. 2019. "Canada Appoints First Ombudsperson Responsible for Corporate Ethics." *The Globe and Mail* (9 April): B.2.

McGill, Hunter, and Lauchlan Munro. 2020. "The Equality Fund: An Experiment Worth Watching." *McLeod Group Blog* (16 April). Accessed 28 October 2021. https://www.mcleodgroup.ca/2020/04/the-equality -fund-an-experiment-worth-watching/.

Meyerhoffer, Sheri. n.d. LinkedIn. Accessed 28 October 2021. https://www .linkedin.com/in/smeyerhoffer/.

Moore, Jen. 2016. Radio interview. Accessed 28 October 2021. https://sound cloud.com/freecityradio/interview-jennifer-moore-from-mining-watch -canada.

– 2018. Latin America Program Coordinator, MiningWatch Canada. Telephone interview by authors, 10 May.

Nanos. 2019. "Latest Reports." Accessed 28 October 2021. http://www.nanos .co/our-insight/.

NRCAN (Natural Resource Canada.) 2018. "Canadian Mining Assets." Accessed 28 October 2021. https://www.nrcan.gc.ca/mining-materials /publications/19323.

Nutt, Samantha. 2012. "The New Humanitarians: Should NGOs Take the Corporate Bait?" *The Globe and Mail* (25 January). Accessed 25 November 2021. http://www.theglobeandmail.com/commentary/should-ngos-take -the-corporate-bait/article1359759/.

O'Neil, Peter. 2012. "Vancouver Environmental Group Accused of 'Laundering' by Oilsands Lobbyists." *Vancouver Sun* (8 August). Accessed 28 October 2021. https://vancouversun.com/news/metro/vancouver -environmental-group-accused-of-laundering-by-oilsands-lobbyists.

Ontario Non-Profit Network (ONN). 2016. "CRA Winding Down Political Activities Audit Program for Charities." Accessed 25 November 2021. https://theonn.ca/cra-winding-down-political-activities-audit-program -for-charities/.

Peck, J., and A. Tickell. 2002. "Neoliberalizing Space." *Antipode* 34, no. 3: 380–404.

Porras Ferreyra, Jaime. 2016. "Justin Trudeau and the Sludge of Canadian Mining Companies." *The New York Times* (3 November 3). Accessed 28 October 2021. https://www.nytimes.com/2016/11/04/opinion/justin-trudeau-and-the-sludge-of-canadian-mining-companies.html.

Ruggie, John. 2008. "Protect, Respect and Remedy: A Framework for Business and Human Rights." Report of the Special Representative of the Secretary-General on the Issue of Human Rights and Transnational Corporations and Other Business Enterprises. UN Doc. A/HRC/8/5. Geneva, Switzerland: United Nations.

– 2011. "Guiding Principles on Business and Human Rights: Implementing the United Nations 'Protect, Respect and Remedy Framework.'" Report of the Special Representative of the Secretary-General on the Issue of Human Rights and Transnational Corporations and Other Business Enterprises, UN Doc. A/HRC/17/31. Geneva, Switzerland: United Nations.

Schreiber, Markus. 2019. "Why Does Justin Trudeau Succumb to Corporate Pressure?" *The Conversation* (5 May). Accessed 28 October 2021. https://theconversation.com/why-does-justin-trudeau-succumb-to-corporate-pressure-116134.

Schulman, Gwendolyn, and Roberto Nieto. 2011. "Foreign Aid to Mining Firms." *The Dominion* (19 December). Accessed 28 October 2021. http://www.dominionpaper.ca/articles/4300.

Simard, Nicolas, and Taj Kudhail. 2018. "Canada without Poverty v. Attorney General of Canada." Fasken's TaxEd International (11 August). Accessed 28 October 2021. http://taxedinternational.com/canada-without-poverty-v-attorney-general-of-canada/?utm_source=Mondaq&utm_medium=syndication&utm_campaign=View-Original.

Tiessen, Rebecca, and Krysten Carrier. 2015. "The Erasure of 'Gender' in Canadian Foreign Policy under the Harper Conservatives: The Significance of the Discursive Shift from "Gender Equality" to "Equality between Women and Men." *Canadian Foreign Policy Journal* 21, no. 2: 95–111.

Wallace, Tina. 2004. "NGO Dilemmas: Trojan Horses for Global Neoliberalism?" *Socialist Register* 40: 202–19.

Women's Rights Policy Group and the Canadian Council for International Cooperation (WRPG/CCIC). 2020. "An Analysis of Civil Society Organizations' Experiences with the Women's Voice and Leadership Program." Accessed 28 October 2021. https://cooperation.ca/wp-content/uploads/2020/07/WVL-report.pdf..

Woroniuk, Beth. 2018. Policy Lead. MATCH International Women's Fund. Telephone interview by authors Kalowatie Deonandan and Toveli Schmuland, 5 May.

8 Mexico–Canada Relations and the Impact of the NAFTA Renegotiations

MARÍA TERESA GUTIÉRREZ HACES[1]

It is difficult to evaluate Canada's evolving presence in Latin America without considering its complex relationship with Mexico, within the North American region.

The clear predominance of the United States in North America has strongly influenced the nature of the relations between Mexico, the United States, and Canada. For over two centuries, the hegemonic character of the United States and its efforts to divide and conquer its subordinate partners have resulted in a fragmentation of the North American region. The literature on North American relations has rarely examined how and under what circumstances the nature of both the Canadian and Mexican states and the relations between them has changed because of their vicinity with the United States. US politics have deeply shaped the behaviour of these two governments, but the two countries have also been able to influence, constrain, dissent, and modify the behaviour of the US government, although to a lesser extent.

In this chapter, I seek to move beyond a separate analysis of bilateral relationships in North America to explain how the two geographical semi-peripheries of the United States have in some cases responded in this vicinity to the United States in a similar fashion, and sometimes differently. This exploration of the differences in the two countries' responses reveals the importance of moving beyond a fatalistic approach to relations within North America that assumes the inevitable dominance of the superpower over its two neighbours and highlights the potential to rebuild the region in more constructive and pragmatic terms.

To do that, this chapter focuses on two elements: One is the renegotiation of the North American Free Trade Agreement (NAFTA) between 2017 and 2019, which undoubtedly modified the character and dynamics of the intertwined relations among the three countries.

This transformation not only changed the balance of power between the United States, Canada, and Mexico but also opened new linkages between Mexicans and Canadians who had not worked together until then, partly due to the US interference but also due to their own interests in forging a preferential relationship with Washington that excluded one another (Gutiérrez Haces 2015). The other focus is to show how in recent years Mexico and Canada have been building their own bilateral relationship within North America, in which they are capable of dissenting with, constraining, challenging, and leveraging their shared neighbour's decisions.

With this analysis, I seek to demonstrate how the development of that Mexico–Canada relationship changed the pre-established equation in the North American region, as demonstrated in the events around the renegotiation of NAFTA and the creation of the new US–Mexico–Canada Agreement (USMCA). Overall, I show how Canada and Mexico were able to develop stronger bilateral economic relations and were more closely linked in regional value chains after the signing of NAFTA. This relationship also shaped their relationship during the USMCA negotiations. The analysis also reveals the importance of the corporate sector in the outcome of the negotiations, as well as the effectiveness of US President Trump's bullying tactics. While the final version of the USMCA included achievements, failures, and many mixed results for all parties, the Mexico–Canada relationship has, on balance, become stronger, moving beyond the comfort zone that NAFTA had created for them and sets the conditions for greater bilateral cooperation in the next period.

Mexico's Place in Canada's Relationship with Latin America

The accelerated rapprochement that has occurred between Canada and Mexico since 1990 cannot be fully understood without a consideration of the dual identity of Mexico as both a Latin American country and a North American country. As discussed in the introduction to this volume, the presence of Canada in Latin America and the Caribbean, including Mexico, was weak, sporadic, and inconstant until the 1970s. Canadian and Latin American authors have extensively analysed this feature for over thirty years, hence my intention in this section is not to elaborate on this aspect, but rather to explain how the Mexico–Canada relationship has influenced Latin American dynamics since the implementation of the North American Free Trade Agreement (1994).

By the 1970s Latin America had undoubtedly emerged as an important alternative for Canada to counterbalance the US dominance. Prime

Minister Pierre Trudeau's Third Option, which advocated for a diversification of trade relations away from the US market, thus represented an unprecedented opportunity to intensify relations with Latin America. The decision of the government of Progressive Conservative Prime Minister Brian Mulroney to pursue a free trade deal with the United States seemed to eliminate the pro–Latin American option. However, after the NAFTA agreement was implemented, Canada decided to move ahead with establishing closer ties with Latin America during the 1990s, using its relationship with Mexico as a template.

Canada took full advantage of the temporary paralysis in the United States caused by the US Congress's rejection of fast-track approval to the signing of new free trade agreements during Bill Clinton's presidency.[2] This temporary commercial advantage led to a transformation of Canada's traditional low-profile commercial diplomacy in Latin America towards a more proactive strategy. This resulted in Canada's signing several bilateral free trade agreements with Latin American countries, as well as trade cooperation and foreign investment protection agreements, based on the provisions contained in NAFTA and in the ill-fated Multilateral Agreement on Investment (MAI) from 1998.[3]

Canada's commercial strategy towards Latin America was selective and strategic, promoting the interests of the most powerful Canadian companies. It consisted of identifying specific commercial and investment niches in which these companies had significant advantages. This strategy reflected Canadian corporate pressures that demanded new forms of government activity to promote their interests in the face of US competition in the Canadian market.

Canada's economic presence in Latin America was not and has not been without difficulties. On the one hand, it is well known that Canada's trade diplomacy primarily promotes the interests of the private sector and that it is considered that a primary function of any government, regardless of its political affiliation, is to facilitate the activities of corporations within and outside the country. From this perspective, embassies and consulates are considered trade promoters. Investors and entrepreneurs mistakenly believe that the signing of an agreement gives them a quasi-monopoly on that market and on the production in each economic sector. For example, when the Mexican government launched a bid for tenders for the construction of subway cars for the cities of Guadalajara and Monterrey, the tender favoured a Spanish company over the bid presented by the Canadian Bombardier-Concarril. This resulted in a conflict between the Canadian ambassador and the Mexican foreign ministry, which led to the ambassador's recall to Ottawa and his subsequent removal from his position in Mexico.

A similar case in Brazil caused a legal complaint between Bombardier and the Brazilian government regarding the Brazilian Aeronautics Company (Embraer), which led the two countries to settle the dispute in the WTO (Mandel-Campbell 2008, 176–81).

The Evolution of Mexico–Canada Economic Relations since NAFTA

An element that is generally overlooked in the contemporary analysis of the bilateral relationship is the fact that while Canada began adopting liberal trade policies since the 1940s, Mexico maintained protectionist economic policies that encouraged import substitution industrialization and that maintained substantial control over foreign investment for decades. This helps explain Canada's earlier minimal economic interest in a country that rejected free trade. Mexico's entry to the General Agreement on Tariffs and Trade (GATT) in 1986, however, gradually opened up the Mexican economy. Later, NAFTA gave Canada a sort of free pass to the Mexican economy, an opportunity from which it profited well, as evidenced by its significant participation in the aerospace industry, automotive production, and the Mexican mining sector.

The implementation and consolidation of NAFTA notably energized the bilateral relationship between Canada and Mexico. The economic data reflect this growing process of consolidation of the Mexico–Canada relationship.

As the data in Figure 8.1 shows, since 1994 (when NAFTA entered into effect), trade flows between Mexico and Canada grew approximately tenfold (Secretariat of Economy, April 2018). It is important to note the bilateral trade balance, which has long been positive for Mexico, even before NAFTA's implementation, except during the period from 1998 to 2008, when the regulatory changes contained in NAFTA temporarily favoured the entry of certain Canadian imports[4] (Gutiérrez Haces 2014, 262–4). According to a report published by the office of the Secretariat of Economy of Mexico in Canada, Mexico's trade surplus grew 0.9 per cent, going from US$13,286 million in January–September 2017 to US$13,408 million in January–September 2018. Mexico remained Canada's third commercial partner with 4.3 per cent of the market, following the United States (61.1 per cent) and China (8.3 per cent).

Mexican multinationals have also expanded their presence in the Canadian economy as displayed by a substantial number of purchases of Canadian companies, especially in the food and construction sectors. For example, the Mexican bakery company BIMBO has made

Figure 8.1. Mexico's Trade Balance with Canada, 1993–2020 (US$ millions)

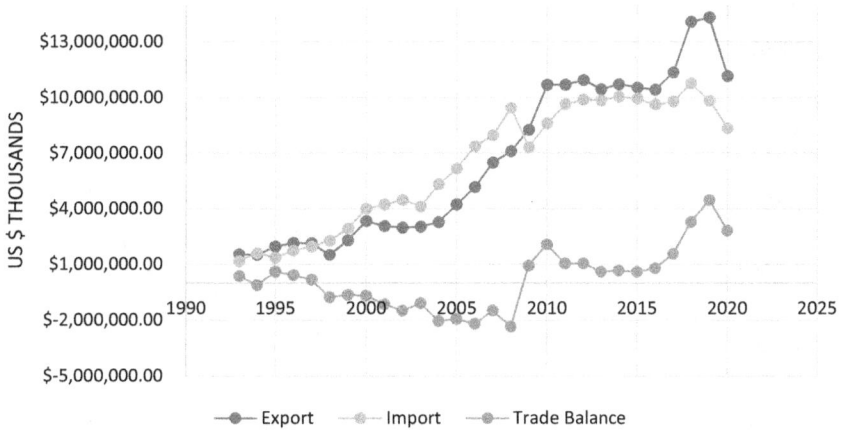

Source: SAT, SE, BANXICO, INEGI. Balanza Comercial de Mercancías de México (1993–2020). SNIEG. Información de Interés Nacional.

Figure 8.2. Canada's Trade Flows and Balance with Mexico, 1997–2020 (US$ millions)

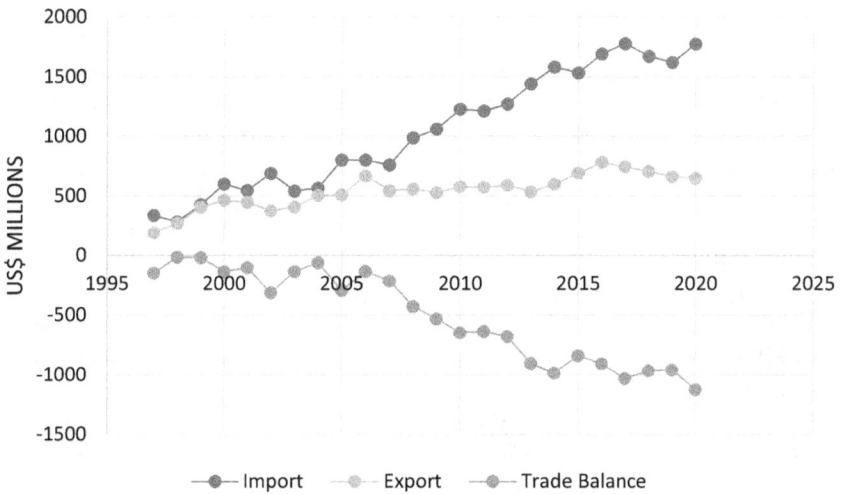

Source: Statistics Canada. Table 12-10-0011-01 International merchandise trade for all countries and by principal trading partners, monthly.

Figure 8.3. Canadian Direct Investment Position in Mexico, 1987–2018 (US$ millions)

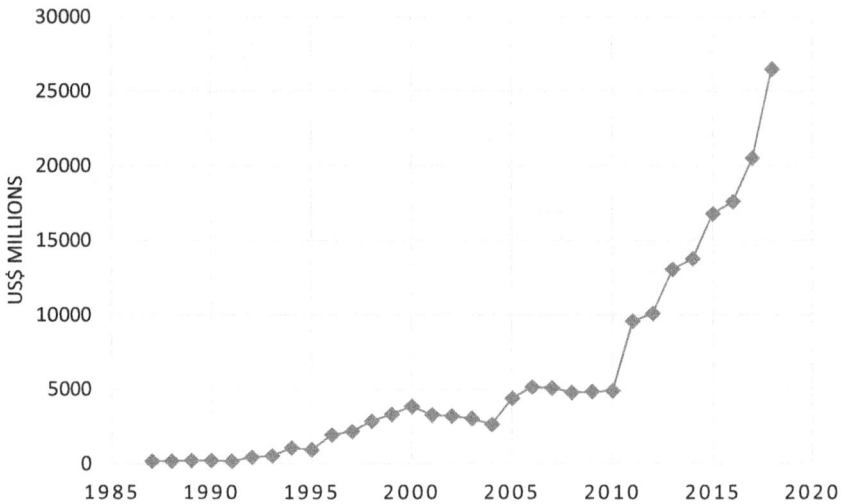

Source: Statistics Canada. Table 36-10-0008-01. International investment position, Canadian direct investment abroad and foreign direct investment in Canada, by country, annual.

substantial acquisitions of former Canadian bakeries, including Canada Bread, Vachon, and Saputo. Mexican companies have also acquired factories and distribution facilities in the aluminum, steel, glass, and alcohol production sectors (Basave and Gutiérrez Haces 2017; 2018). The growing Mexican investment in Canada is an interesting example of corporate integration in North America. Mexican and Canadian companies increasingly tend to be interconnected in global value chains, especially in the automotive, aerospace, oil, and mining sectors.

As well, since NAFTA's implementation, Canada has been a key player in the regional integration of productive sectors through foreign direct investment (FDI) in Mexico. From January 1999 to June 2017, Canadian FDI in Mexico amounted to US$29,675 million (Secretariat of Economy, November 2017, 5), which meant Canada ranked fourth among foreign investors in the country. For example, there are around 60 Canadian manufacturing companies in the automotive sector, with 115 factories across Mexico.

In sum, 43.9 per cent of Canada's total investment went to mining, of which 98.7 per cent focused on metal and non-metal minerals

Figure 8.4. Canadian FDI in Mexico by Destination Sector (1999–2018) (US$ millions)

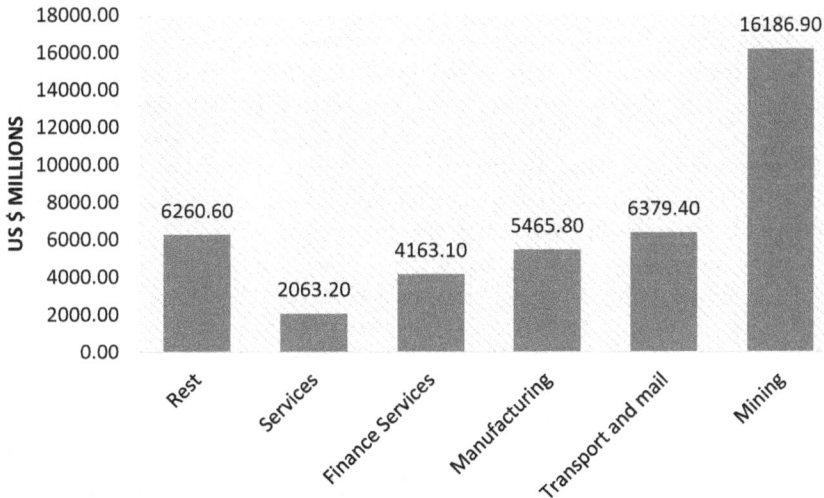

Source: Secretariat of Economy, "Direct Investment from Canada to Mexico," [online], Mexico, February 2018, URL: https: //www.gob.mx/cms/uploads/attachment/file/302801 /Distribucion_de_inversion_de_Canad_.pdf.

(Secretariat of Economy 2015); 54.6 per cent of Canadian mining investment is concentrated in five Mexican states: Zacatecas, Chihuahua, Mexico City, Sinaloa, and Coahuila (ibid.). In September 2018 there were 3,944 mining companies with Canadian capital participation investing in Mexico.

With NAFTA, the United States achieved a privileged position with Mexico and remains Mexico's first trade partner.[5] Nevertheless, Canada has an important leverage in relation to FDI in Mexico, primarily due to the investment made in Mexican mining companies. According to Pro México, a Mexican government agency devoted to promoting foreign investment in Mexico, by the end of Peña Nieto's administration (2018), 74 per cent of the FDI in that sector came from Canadian capital, with 200 Canadian mining companies investing in over 942 projects (20 Minutos 2017).

The intensification of Canadian mining investment in Mexico has resulted in the greater institutionalization of the Canada–Mexico relationship. The two countries have created a Mining Working Group to foster mining and environmental development. As a product of their work, during the visit of President Enrique Peña Nieto to Canada in June 2016, a joint statement was signed on environmental cooperation between Canada and Mexico to include Indigenous issues, along with economic development, environmental regulation, and technology (Canada-Mexico Partnership 2016). An MOU (memorandum of understanding) was also reached that included the exchange of information and knowledge as well as technical expertise between Canada's Department of Natural Resources and Mexico's Secretariat of Economy; periodic meetings between government representatives; and joint consultations with industry, academics, and other stakeholder groups (Natural Resources Canada 2017).

Additionally, the mining group agreed to create a platform to share positions and strengthen their approaches to the Asia-Pacific Economic Cooperation (APEC), the Trans-Pacific Partnership (TPP), and the Organization for Economic Cooperation and Development (OECD). During the 2016 meeting of the working group in 2016, an MOU between Natural Resources Canada and the Secretariat of Economy was signed regarding collaboration in Sustainable Mineral Resource Development (Natural Resources Canada 2017). For future activities, the mining group intends to compile a case study of both Mexico and Canada to share environmental technologies. The priority of this project will be to find the best way for mine closure and remediation. Another important and innovative project is led by the Mexican Geological Service and the United States Geological Survey that also involves the Canadian minerals and metals sectors: "Net Import Value of North America (without Fuel): Minerals and Metals that Are Critical to (or Allow) Clean Energy Technology" (Canada-Mexico Partnership 2016). Also, during 2016, with regard to Mexico's Mining Promotion Trust (FIFOMI, in Spanish), Mexico and Canada agreed to explore the possibility of promoting a larger number of mining projects in Mexico in the future.

During the first three years in office of Andrés Manuel López Obrador (December 2018– December 2021), domestic and foreign extractive companies faced an extremely complex situation in terms of the future of their investments. The new Mexican government has given contradictory signals regarding the investment and operations of mining, electricity, and hydrocarbons companies and therefore the official stance is still unclear. In particular, the Fund for Regional Sustainable Development of Mining States and Municipalities, one of the most visible accomplishments of the 2013 Constitutional reform on energy[6] collected a new tax that represents 7.5 per cent of the annual profits of all mining companies, plus 0.5 per cent on extraction of gold, silver, and platinum. This fund, destined to public works in communities affected by mining, will no longer be destined to such activity and will be managed directly by the federation rather than by the state and municipal governments, contradicting López Obrador's promises when he took office, when he stated support would go to the affected communities.

This change, along with others of the same kind, have caused a radical change in the orientation and structure of the incoming government. One of its first actions in 2019, pointed to the disappearance of more than one hundred government trusts (*Fideicomiso* in Spanish), considering that it was necessary to redirect the use of government resources towards greater austerity. In reality, these changes have been aimed at obtaining sufficient resources to finance its star projects, such as the construction of the Dos Bocas oil refinery, the Mayan train, the Santa Lucía airport, and promoting their new welfare programmes.

As a result of these initiatives, it was decided to change the nomenclature from the Mining Fund to a Fund for the Development of Mining Areas. This change meant much more than just the name; it implied upsetting the original nature of the Mining Fund by altering its central objective: serving the needs of municipalities and mining communities and ignoring their rights over resource management by eliminating participation of its inhabitants in defining projects and supervising the use of resources.

The creation of the new fund was carried out with the consent of the federal government through the publication in the *Official Gazette of the Federation* of the guidelines that establish the application of the resources of the newborn fund. Among others, it eliminates the provision that links the application of the fund to communities and municipalities where mining extraction occurs and the mechanism for its execution is centred in the Ministry of Education. Until December 2018, the income obtained from the payment of rights, of which 77.5 per cent was assigned to the Mining Fund, 2.5 per cent to the Secretary of Agrarian, Territorial and Urban Development (SEDATU in Spanish) for the operation of the fund and, 20 per cent was channeled to the Ministry of Finance, for infrastructure works.

Since most foreign companies in the mining sector are Canadian, this measure has to some extent soured the relationship between the two governments and with the municipalities affected. The first Mining Fund was an innovative initiative adopted by the federal government in 2013. For the first time, it was recognized that part of the taxes collected from mining must favour those municipalities where such activity takes place. At first, mining companies disagreed with this change, but in the long run they recognized that it was an important measure to support social development and political stability in those towns. From 2014 to 2018, the Fund amounted to $11,369 million pesos (Fondo para

Figure 8.5. Tax Contributions to the Mining Fund (2014–17) (MX$ thousand million)

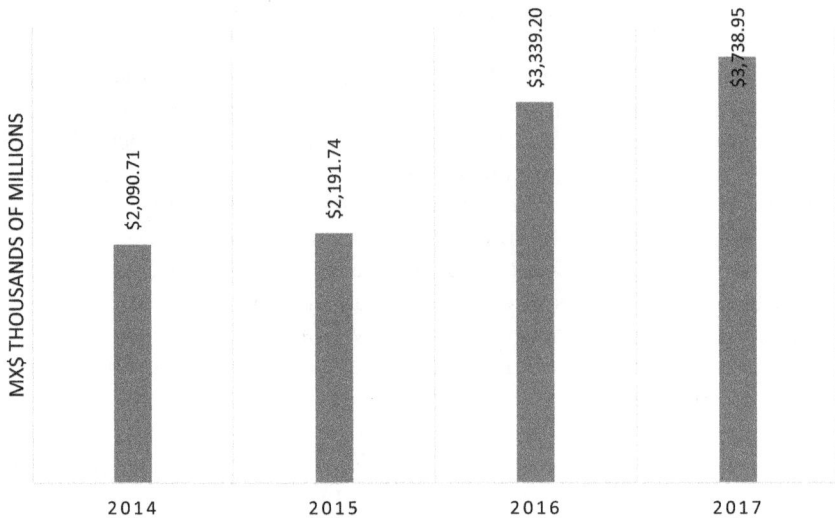

Source: Own elaboration with data of Mining Fund, URL: https://www.gob.mx/sedatu /acciones-y programas/fondo-minero-para-el-desarrollo-regional-sustentable.

el Desarrollo Regional Sustentable de Estados y Municipios Mineros, SEDATU, 2018).

Overall, the priorities of Mexico and Canada in bilateral cooperation were not altered by the USMCA negotiations and continued being a priority until the end of 2018 with the conclusion of Enrique Peña Nieto's presidency. Upon the arrival of the new Mexican government of Andrés Manuel López Obrador, the governance of mineral resources, trade, investment, and access to markets as well as the exchange of information, corporate social responsibility, socio-environmental issues, technology and innovation, and cooperation in multilateral forums are undergoing a period of uncertainty.

Canada and Mexico in the Oil Sector

In recent years, North American governments and corporations have viewed the oil sector as one area for future cooperation thanks to Mexico's opening in this sector in 2013. During the bidding process, Canadian company Renaissance Oil obtained four fields for exploration. This is the first Canadian private company that will produce oil in Mexico (ibid.).

An issue to which the oil industry leaders gave a lot of attention during the NAFTA renegotiation was how NAFTA's Chapter 11 rules would be applied in their sector. Over 750 companies that comprise the Mexican Association of Hydrocarbon Companies (AMEXHI), the American Petroleum Institute (API), and the Canadian Association of Petroleum Producers (CAPP), reaffirmed in 2018 that they supported a modernized NAFTA that contributed to the success of the North American energy market. They also added that if it were not possible for the three countries to agree on such a modernized agreement, they urged the relevant policymakers to maintain NAFTA because of its considerable mutual benefits in the energy sector which would decrease dramatically without a trilateral agreement. The association continued to reject proposals such as Trump's original demand for a "sunset clause," which would have required the new agreement to be renewed every five years, with sunset as the default option if an agreement were not reached. Another proposal they opposed was a "voluntary inclusion/exclusion" Investor State Dispute Settlement (ISDS) mechanism which would weaken the stability and protection of long-term investments (AMEXHI 2017).

Analysing the results of the renegotiation, the oil sector was one of the biggest winners in USMCA as the sunset clause proposed by the US negotiators was replaced with a requirement that the deal be renewed after sixteen years. As well, ISDS was maintained between Mexico and the United States in the oil and gas sector. In fact, a new clause was added that gives companies the right to claim damages due to investments being affected under two modes: First, regarding a party's breach of obligations under National Treatment, Most Favoured Nation, and Direct Expropriation, and secondly when an investor is part of a contract in sectors such as hydrocarbons and gas, telecommunications, energy generation, transport, and infrastructure projects.

On the other hand, Canada scored two major triumphs. The first was the elimination of ISDS for conflicts concerning investment between the United States and Canada. It is worthwhile noting that this does not prevent a company from suing either Mexico or Canada under the rules of any other agreement such as the TPP and CETA. The other was the disappearance of the "energy proportionality" obligation in NAFTA that required Canada to maintain a consistent share of energy exports to the United States as a proportion of domestic supply based on a three-year average, in case of a crisis or conflict.

The fact that the same agreement embraces two diametrically opposed positions reflects, on the one hand, the degree of autonomy and maturity of both countries, not only in what concerns the interests of the United States but also with reference to their mutual relations.

It is understandable that Canadians celebrated the abolition of the energy proportionality clause in NAFTA. The reason why the US negotiators consented to this must be attributed to various factors. First and foremost, the United States currently depends much less on Canadian and Mexican oil supplies due to the growth in oil production inside their territory and the use of non-traditional extraction methods, such as fracking.

As well, the energy reform to the Mexican constitution in 2013 opened extractive activity in the hydrocarbon sector to Canadian and US investors, though not exclusively.[7] The activity in this sector is not constrained to oil exploration and extraction; it also represents a lucrative business for the service sector in both countries. US investment in the energy sector comes first, while Canada is ranked fifth, after Colombia and the United Kingdom.

The American Petroleum Institute (API), the Canadian Association of Petroleum Producers (CAPP), and the Mexican Association of Hydrocarbon Companies (AMEXHI) supported two fundamental demands in NAFTA's renegotiation: the reduction and elimination of tariffs on energy products and the preservation of a dispute-resolution mechanism between investors and states to prevent a state from expropriating assets without compensation. The president of AMEXHI at that time, Alberto de la Fuente, stated: "After Mexico's energy reform, NAFTA attracted a great deal of investments, developed infrastructure, facilitated a more intense trade exchange. The synergy between NAFTA and the energy reform in Mexico is essential to attract investments, develop integrated value chains, and increase competitiveness in North America" (Serrano 2017).

When Mexico proposed adding a chapter on energy to the new NAFTA, Canada agreed. In contrast, US investors asked their government to use the mechanism known as the ratchet, which would prevent the implemented reforms from being reversed (Morales 2017). At COP22, Canada announced an investment of $7 million to support the reduction of climate pollutants, such as methane, in the oil and gas sector over the next four years (Canada-Mexico Partnership 2016). Canada has emphatically supported Mexico's energy sector in various fronts, such as regulatory cooperation. Canada's regulatory organizations have worked actively with their Mexican counterparts to share better practices. For example, the University of Calgary and the University of Alberta received $55 million through SENER (Secretariat of Energy) and CONACYT (National Council for Science and Technology) for academic exchanges and the development of an array of technical and postgraduate programmes in energy, oil, gas, and research and development issues, seeking to foster long-term relations between the two countries.

Mexico–Canada Partnerships Prior to NAFTA Renegotiation

For the remaining issues, the most important forum for bilateral cooperation is the Mexico-Canada Partnership that was established in 2004, which sought to boost the bilateral relationship and participation of various governments, scholars, and businesspeople, through various working groups, including agribusiness, forestry, human capital, trade, investment and innovation, energy, the environment, and labour mobility (Secretariat of Foreign Affairs 2016). A group on mining, discussed above, was also established in 2015 and held its first meeting in 2016.

During the annual meetings held by these groups, an issue that has been addressed repeatedly has been Mexico's energy reform, given the immense possibilities it has opened up to Canadian investors. In June 2014 Alberta Energy Regulator signed an MOU with Mexico's National Commission on Hydrocarbons on best regulatory practices, besides the technological exchange in the sector. Currently, it is working with Mexico to develop a Centre of Regulatory Excellence, to train regulators at the international level (ibid.). In addition to this, Ontario's Electric System and Mexico's National Center on Energy Control have united to facilitate technical exchanges in the electricity market (ibid.). In the same vein, Mexico entered into an MOU with Manitoba Hydro International Ltd. to increase cooperation between both countries and exchange experiences on cutting-edge technologies and best practices in the development of renewable energy projects (Secretariat of Energy 2015).

From the perspective of both governments, one of the most successful projects is the Canada–Mexico Seasonal Agricultural Workers' Program that started in 1974 with the execution of an MOU between the two governments. According to the Mexican Secretariat of Foreign Affairs, this programme has employed 320,000 workers since its creation (Secretariat of Foreign Affairs 2016). According to this MOU, the Mexican workers participating in the programme are entitled to have dignified housing, a previously agreed wage, labour guarantees, and access to provincial medical insurance during their stay in Canada (Consulate General of Mexico in Montreal). Nonetheless, although the programme has advanced substantially, it has received enormous criticism both in Mexico and Canada based on concerns around labour and human rights (Vanegas 2019; Gabriel and Macdonald; 2011; 2012).

Another area of cooperation promoted under the CMP is in tourism. Over 1.7 million Canadians visited Mexico in 2015, while just over

180,000 Mexicans travelled to Canada. After having withdrawn the obligation for Mexicans to have a visa to travel to Canada (2004), an electronic travel authorization (eTA) was implemented to control the entry to Canadian territory of Mexicans who request it (Secretariat of Foreign Affairs 2016).

With regard to cooperation in education, various agreements have promoted academic mobility between students and researchers, such as the agreement between the Secretariat of Public Education and the University of British Columbia, in addition to the *Proyecta 10,000* programme that up to 2015 benefited 2,991 students (ibid.).

Mexico–Canada Relations during the USMCA Negotiations

The relations between Mexico and Canada took an unexpected turn in 2017. Both countries were practically forced to engage in a NAFTA renegotiation process for which they had not been looking. The renegotiation was imposed unilaterally by the incoming president of the United States, Donald Trump, who repeatedly stated during his campaign that if he made it to the presidency he would formally withdraw from NAFTA, as he considered it was harmful to the US economy and its citizens.

This announcement placed the Canada–Mexico relationship in a sort of impasse, wherein each of the two governments started a unilateral process of negotiation, which in the long run almost paralysed the bilateral linkages that had been established under NAFTA. Canada went through this process assuming that its special relationship with the United States would give, if not the same privileges as before, at least certain preference in its engagement with the United States. In contrast, Mexico entered the negotiations in an extremely fragile position due to Trump's unceasing attacks on Mexico regarding the Mexican government's inability to stop the flow of undocumented migration into the United States, and his demand for Mexico to pay for a wall on the US–Mexico border.

On 25 January 2017 the Canadian negotiating team jumpstarted the talks with a strategy marked by sharp reverses, at times stating that Canada would leave Mexico on its own in the NAFTA renegotiations, while, at other times, asserting that Canada and Mexico would face the renegotiation together. On the other hand, the Mexican negotiating team officially maintained, from the beginning to the end, the need to advance on the process together.

Like in the first NAFTA negotiations (1992–4), once again, the labour conditions governing the three countries represented a polarizing factor at the start of the renegotiations. The AFL-CIO requested that a modernized NAFTA must guarantee all workers, irrespective of the sector they worked in, a right to receive a living wage (Riquelme 2017). It also demanded that exporting a product where workers at any point of the productive chain may have received a smaller wage would mean a breach of the agreement (ibid.). Jerry Dias, National President of Canada's largest private-sector union, Unifor, held: "[Mexico] somehow has to keep its people in poverty to create employment [which] is nonsense and outrageous" (Fariza 2017). In response, the Mexican government made it clear that it would not accept the inclusion of wage issues in the renegotiated NAFTA. In fact, Mexican businesspeople stated: "Mexico cannot interfere in the labour issues of the US or Canada. And we ask them the same; do not interfere in these issues" (Vanguardia MX 2017).

Throughout a year and a half of negotiations, the issue of wages came up intermittently, as a spearhead to weaken the Mexican negotiators and pressure them to give in on some aspects, such as determining what the percentage of the regional rules of origin should be. In contrast, NAFTA's Chapter 19 concerning state-to-state dispute resolution was an area of convergence between Mexico and Canada, as both opposed US arguments seeking to eliminate it. On several occasions, Canada stated either alone or together with Mexico that one of its goals was to protect the dispute-resolution mechanism included in the original NAFTA (Riquelme 2017), which represented continuity in the Canadian strategy since the negotiation of its first free trade agreement with the United States in 1988.

For over a year, President Trump focused on the need to rebalance the US trade deficit, based on at least three proposals. First, he sought to end the dispute-settlement mechanisms on trade and investment, which he claimed were partly accountable for causing the deficit; secondly, he wanted to establish new rules of origin, demanding that more than 50 per cent of production take place on US soil; and thirdly, the automotive sector was used as an arm-twisting measure, directly attacking the structure of its value chain, one of NAFTA's most integrated. As a corollary, months later, Trump decided to reinforce his harassment strategy by imposing a tariff on steel and aluminum imports on both Canada and Mexico.

In January 2018, Mexico and Canada jointly rejected the US government's proposal to modify the Chapter 11 ISDS mechanism (Noticieros Televisa 2018). According to the *Wall Street Journal*, the US position was that resorting to a dispute-resolution panel should be optional since, in the eyes of the US government, it harmed its sovereignty by allowing multinational companies to avoid US courts (ibid.). Concomitantly, the Mexican and Canadian authorities held that they preferred to eliminate the agreement's arbitration panels and create their own bilateral pact, instead of being part of a system embedded in NAFTA where different countries have different rights (ibid.). This stance was practically dismissed on the last day of negotiations, when the Canadian negotiating team decided to exclude itself from the arbitration mechanism on investment and exclusively abide by Chapter 19's trade arbitration. In contrast, Mexico had already accepted weeks before, during its own negotiation with the United States, the two arbitration mechanisms to settle disputes with the United States, except that the one on investment would be used only in disputes over oil, infrastructure, and telecommunications, leaving out the rest of the economic sectors, such as mining and manufacturing. This decision will directly affect Canadian investment in Mexico, which for more than twenty years has focused on mining activity.

Paradoxically, as the negotiations were moving forward, US businesses decided to support Mexico and Canada's posture in order not to lose the agreement and to maintain the dispute settlement chapters contained in NAFTA, especially the one on investment. The US Chamber of Commerce established a framework with four items that summarized what they expected from the NAFTA renegotiation: 1) doing no harm, not interrupting the annual trade that generated $1.3 trillion, or 14 million US jobs, and not affecting the 125,000 small and medium-sized US enterprises; 2) amending NAFTA instead of ending it; 3) maintaining the trilateral nature of the agreement; and 4) advancing promptly with the renegotiation since uncertainty on the future exchanges of the United States with Canada and Mexico would suppress economic growth in both countries. In this regard, Mexican businesspeople, through the Business Coordinating Council (CCE), simply rejected the US proposal, stating: "We want an FTA that allows us to continue being the most competitive region in the world, but not at all costs."

The corporate sector received vigorous support from a group of US senators who sent a letter to President Trump in which they reminded him of the multiple benefits NAFTA had brought about and what this represented in numbers. In addition, they included their expectations for the renegotiations that mirrored the US corporate sector's demands: modernizing NAFTA, not ending it. In the context of defending NAFTA, the Mexican private sector lobbied US and Canadian businesspeople before the fifth round of negotiations "to get together and counterbalance any possibility of the US pulling out of the agreement," as asserted

Moisés Kalach, coordinator of the "side door" at the Business Summit Mexico 2017 (Patiño 2017).

Again, in March 2018, a statement made by the president of Unifor caused confusion and concern when he claimed that the United States was hindering the negotiations and therefore suggested that Mexico and Canada should leave the agreement (Esquivel 2018b). He also stated: "You would expect a commercial partner who wants to reach an agreement with Canada and Mexico would do everything he can, but the US has put up all the obstacles" (ibid.). These statements came after Trump announced the application of the tariff on steel and aluminum imports (ibid.).

Trump was not the only head of state pressured by the business sector. Mexican president Peña Nieto also received an unusual letter in 2017 signed by the Canadian Chamber of Commerce (CANCHAM) in Mexico. In it, they asked him to stand by NAFTA and its modernization. In addition, the CANCHAM president stated, "For over 22 years, NAFTA had achieved between Canada and Mexico, not only an exponential growth in international trade between them, but also an institutional framework that had given certainty for investments to flourish, especially with Mexican companies of Canadian capital" (Saldaña 2017). He also pointed out that he agreed to defend NAFTA and its modernization to include new areas of trade (ibid.).

To provide even greater support to the renegotiation process, on 24 January 2017 thirty-two organizations from the three countries agreed to demand that their respective governments maintain the free trade agreement and reinforce it to preserve employment (Esquivel 2018a). Ricardo Navarro, representative of Concanaco, asserted: "Our request is to reinforce it, update it, and maintain access to the US market" (ibid.).

The NAFTA renegotiations not only brought responses from business and federal politicians. In fact, governors of the US border states pushed President Trump openly, before an imminent danger of losing very profitable businesses with Mexico and Canada that were protected by NAFTA. For example, the governor of Texas, Greg Abbott, said in an open letter to Trump that it was crucial to maintain the dispute-resolution mechanism on investment since its principles reflected the fundamental rights of protecting private property under US law, including justice, due process, non-discrimination, and compensation for expropriation. Abbott added that after the 2013 energy reform in Mexico, a series of opportunities had opened up for investors, which US companies had greatly profited from thanks to the protection NAFTA provided.

The strengthening of Mexico–Canada relations during the renegotiation was also displayed in the strategy implemented by the two countries

when addressing the rules of origin they had to contemplate in the automotive and textile sectors. Both countries sustained that they would
maintain a joint position in opposition to the United States' initial proposal on the rule of origin for the automotive sector, which proposed to
increase the regional content from 62.5 per cent in NAFTA to 85 per cent
(La Silla Rota 2017). Over a year and a half of negotiations, Canada and
Mexico decided to look beyond Trump's intransigence. The fact that
they both participated in the launch of the Trans-Pacific Partnership
(TPP), meant that both countries maintained similar positions.

Despite these shared objectives, as part of an unfortunate communications strategy, on 25 January 2017 the Canadian government declared
that it would leave Mexico alone in the NAFTA renegotiation: "We
love our Mexican friends, but our national interests go first, and the
friendship comes second. The two are not mutually exclusive" (Tele Sur
2017). It goes without saying how startling this statement – the first of
some others to come – was to the Mexican negotiating team.

The Canadian statements greatly influenced the decisions that the
Mexican negotiating group would take later. Those who did not closely
follow the negotiation process could legitimately see it as surprising
and even disloyal for the Mexican negotiating group to conclude its
own negotiation without Canadian accompaniment. However, this outcome occurred with the knowledge and consent of the Canadian negotiators, who possibly underestimated the negotiating capacity of the
Mexicans. If the Canadians expressly withdrew from the negotiation
to return to Ottawa, it was because they sought to break the impasse
they were facing. Canadians consulted with the Mexican negotiators on
their position, who, at the same time, decided to continue negotiating
those aspects that were not strictly of bilateral interest. When the Mexicans finally concluded their negotiations and announced a deal on the
whole agreement, the Canadians were at first surprised and displeased,
despite the fact that they had been consulted on the strategy.

Faced with this *fait accompli* the Canadian government preferred to
keep a relative silence, letting social networks and the media criticize
the Mexican negotiators. The momentary withdrawal of Canadian
negotiators, as well as the green light given to Mexicans to continue
their negotiations, were part of a calculated strategy that sought to
pressure Trump and his negotiators with a withdrawal from the negotiating table. Nevertheless, when the United States disclosed the items,
it wished to bring to the table at the renegotiation rounds[8] – especially
the elimination of Chapters 11 and 19 – Canada's Minister of Foreign
Affairs Chrystia Freeland and the Mexican Secretary of Foreign Affairs
Luis Videgaray stated that the agreement had to be trilateral.

Unquestionably, Canada and the United States had important disagreements, particularly during the fourth round of negotiations, when Washington stated it would impose a 220 per cent tariff on Canada's Bombardier's C series aircraft, arguing that they benefit from subsidies to the detriment of US company Boeing (Panorama 2017).

Mexico and Canada maintained a joint position on the chapter that regulates the national content percentage. The rules of origin in the automotive and textile sectors were an issue negotiated at the final stage of the process (La Silla Rota 2017). The Mexican government held: "Same as Canada, we (Mexico), do not agree on the US proposal [to increase the national percentage to 85 per cent because] we consider it unviable" (ibid.). Canada's ambassador in Mexico at that time, Pierre Alarie, asserted that in case the United States withdrew from NAFTA, his country would maintain the agreement with Mexico and it would become a bilateral relation: "We're friends and partners with Mexico: it is a very important partner, we're here to endorse it, we're not leaving it aside" (Cámara de Comercio y Servicios). In this line, members of Canada's House of Commons and the Mexican deputies jointly expressed their decision to support a bilateral trade agreement if NAFTA negotiations failed: "Both of us have agreed that we must go along together, that we must keep sitting at the table, that we must attempt for this agreement to conclude successfully, for the benefit of all the parties" (Pablo 2017).

As we have seen, politicians, governments, and corporations came together in promoting a new trade agreement. The Canadian business sector expressed its interest in maintaining a trilateral NAFTA, arguing that Canada was a reliable partner that would collaborate with Mexico and the United States to reach the best agreement possible for the three countries (ibid.). The Canadian business sector highlighted in various forums the weight and importance of its business interests in Mexico, especially mining. Eventually a final deal was reached.

Content of the USMCA

This chapter has sought to analyse the ups and downs of the relations between Mexico and Canada, produced by the NAFTA renegotiation process. Undoubtedly, this was a difficult process that brought to the table old problems that had persisted since NAFTA's entry into force. Canada tried to maintain a trilateral focus throughout the negotiation; nonetheless, like Mexico, it had to face a devastating attack coming from the US president and from the US negotiating team.

From the beginning, Trump's main weapon against his opponents was to drive them to the point of exhaustion that harmed them both individually and collectively and that diverted investment from both Mexico and Canada to the United States. To this strategy of exhausting his competitors, Trump added chaos, cunningly contradicting himself with the purpose of confusing Mexicans and Canadians. When the negotiators understood this, they decided to disengage from Twitter to stop contaminating the negotiations. However, neither Chrystia Freeland nor Mexican trade minister Ildefonso Guajardo took into consideration the fact that millions of Mexicans and Canadians were forming their opinions on the progress of the negotiations based on the US president's statements, while the negotiation were trapped in a secrecy imposed to a large extent by the US team.

Canada introduced new topics to the negotiation and attempted to raise concerns coming from civil society regarding gender, the environment, and Indigenous rights, but it mostly lost these battles. The agenda of the US government and Canadian corporations took precedent over the inclusion of citizens' demands in the negotiations. Minister Freeland did not calculate the strength of the presence of a cohesive trinational group of businesspeople for whom national identity was largely irrelevant. The corporate interests of North America, after 25 years of NAFTA, were very similar and did not necessarily coincide with her approach.

Accepting a trilateral mechanism that sought to empower surveillance over the exchange rate policy, bypassing the decision of the central bank was a last-minute blow landed by US negotiators, who were aware that at least the Mexican negotiators would be practically out of office from 1 December 2018 with the change of administration and the legislature in Mexico, and therefore they would not provide much opposition.

The war of attrition did its work against Mexicans and Canadians. The negotiators gave in on issues that they considered unacceptable for their countries, like the percentage increase in the regional content rules, which was finally fixed at 75 per cent. This way, it was accepted that a vehicle to be exported within North America would have to be manufactured with a regional content that complied with that proportion. Unquestionably, this decision will affect international suppliers of the automotive industry and may affect the competitiveness of the North American value chain.

Canadian negotiators managed to achieve the inclusion of Chapter 19, which covers the dispute settlement mechanism, in the USMCA, but in return, they surrendered part of the control of their dairy, poultry, and egg market, granting US products access to 3.5 per cent of it, representing $16 billion a year. In response, the Canadian government announced that it would prepare a compensation package for affected farmers.

The USMCA contains at least three very powerful provisions that will disrupt North American relations. The first is the direct involvement of Mexico and Canada in the US trade war against China, since Clause 32.10 of Chapter 32 on Exceptions and General Provisions sets forth that none of the three countries may enter into economic agreements with non-market countries.[9] Perhaps the Mexican and Canadian negotiators thought the clause had a very constrained scope since it only addressed the action of negotiating a free trade agreement, not trading as such. However, this clause limited Mexico's and Canada's commercial futures, dooming them to import and export to China without the preferential conditions that a free trade agreement with this Asian country would offer them.

The second provision is contained in Chapter 33, Clause 33.4 of the chapter on Macroeconomic Policies and Exchange Rate Matters, in which the countries entrust a trilateral Macroeconomic Committee composed by representatives of the three countries some of the major decisions on national exchange-rate policy, setting forth that "each Party should achieve and maintain a market-determined exchange rate regime." This provision erodes the autonomy of the central bank, which will stop intervening in the exchange-rate regime according to the economic circumstances and will stop employing part of its international reserves to rebalance the highs and lows of the exchange rate; in addition, it also loses its capacity to induce moderate devaluations to rebalance foreign trade. In the opinion of Stephanie Segal: "USMCA currency provisions set a new precedent. The trilateral Macroeconomic Committee aims to prevent any USMCA members from pursuing a competitive devaluation" (Segal 2018).

Thirdly, USMCA has two very differentiated approaches with respect to ISDS. It has been totally eliminated for the United States and Canada,[10] but its disappearance will take time, and therefore companies will be able to continue suing the Canadian government (Sinclair 2018); in the case of Mexico and the United States, although the mechanism remained, it contained significant changes. The arbitration mechanism would likely disappear gradually; it no longer covered the entire economic universe and limited the protection to investment of companies in certain activities.

Damages may be claimed due to harmful investment under two scenarios: violating the obligations of national treatment, Most Favoured Nation, and direct expropriation, or breaching a contract in the sectors of hydrocarbons and gas, telecommunications, energy generation, transport, and infrastructure projects. Finally, the dispute-settlement mechanism was updated by incorporating new elements: transparency of the arbitration procedure in both written and oral performances is favoured; acceptance of new rules established by UNCITRAL is allowed; the incorporation of ethical rules that arbitrators will have to observe; swifter procedures to settle jurisdiction matters are contemplated, as well as rules for the participation of non-contesting parties; and termination of arbitration due to procedural inactivity (Secretariat of Economy 2018).

As of today, the analysis of the process and its results offers multiple interpretations. For some, USMCA is an extension of what is already wrong, and for others it somehow means a victory for the Canadian social movements that fought for twenty-five years to see the elimination of the investment arbitration mechanism. Only time will give us the answer.

Postscript

On 29 May 2019 Prime Minister Trudeau presented the USMCA for consideration by the Parliament of Canada. Two days later President López Obrador handed the text to the Senate of the Republic to begin the ratification process. For his part, Robert Lighthizer issued a draft of the Statement of Administrative Act so that in the next thirty days the treaty would be sent to the House of Representatives to be examined. The ratification process began shortly after President Trump decided to eliminate the extraordinary duty that was levied on steel and aluminum exports from his two neighbours.

The Mexican government's rejoicing was very ephemeral, since on 30 May, Trump announced unilaterally, that is, without consulting the Department of Commerce or the State Department, much less the Congress, that as of 10 June, it would impose a 5 per cent tariff on all Mexican exports without exception and that the tariff would gradually increase monthly to 25 per cent if the Mexican government did not show that it was really controlling and preventing illegal migration through the northern border of Mexico.

With this announcement, all economic indicators fell to worrying levels, and the Mexican peso suffered a considerable devaluation. The sales of Mexican multinationals in the Mexican and New York Stock Exchanges fell, producing a chain reaction that directly punished Mexican consumers.

Linking migration issues with bilateral trade is a measure that, in addition to lacking any link with economic logic, brutally violates the human rights of migrants. Trump was determined to use any kind of retaliatory measure to subdue any country without any dissent. Regrettably, it is a fact that even if the control measures were intensified, undocumented Central American migrants would inevitably continue to travel north because they did not have much else to lose, given the situation of violence and extreme poverty that prevailed in their countries of origin.

Throughout the USMCA negotiation process, as well as during its legislative approval and even after its launch in July 2020, the United States did not cease to threaten and unilaterally apply punitive tariffs on merchandise from Mexico and Canada. This situation exceeded the acceptable limits of a strictly commercial war, where the imposition of a tariff, despite the existence of a free trade agreement, became the most used resource by the United States to impose its rules to international trade. Shortly after the USMCA was signed by the three presidents during the G20 Summit in Buenos Aires, pressure increased from the US Congress regarding its demand that Mexico should change its labour laws and the criteria for determining the minimum wage, especially in the automotive sector. Finally, the new labour law was approved, and it was hoped that with this initiative, American demands would decline, but Trump's decision to impose an incremental tariff on all Mexican exports proved that Mexico's relationship with the United States continued to be erratic and chaotic for both countries.

In August 2020, the US president once again threatened Canada with a 10 per cent tax on its aluminium exports, which would affect both countries' economies, given the increase in prices on final goods produced with that product. Unlike the retaliatory measures against the unilateral imposition of a tariff by the United States, invoking Clause 232 on security, under the new USMCA rules, neither Canada nor Mexico, would be able to make use of the so-called mechanism carousel which in the past allowed them to rotate products with tariffs to promote uncertainty in the United States. Simultaneously, the US government's Inter-Institutional Labor Committee for Monitoring and Enforcement set up a USMCA hotline to receive confidential information sent by interested parties on labour issues between the two countries at the US embassy in Mexico, a mechanism many Mexicans saw as representing undue interference in their country's affairs.

The bilateral US–Mexico agreement also contained a Rapid Response Labor Mechanism, the first of its kind that allowed the United States to take enforcement measures against Mexican factories if they did not comply with Mexican freedom of association and collective bargaining laws. The USMCA already had a budget of $210 million for implementation activities, of which $180 million would be used for four years for technical assistance projects related to the treaty and $30 million for eight years for training and monitoring.

In July 2020, the USMCA finally came into force. However, the event was totally overshadowed by the COVID-19 crisis. The pandemic by its very nature produced a chain reaction that affected all the value chains that linked the economies of North America. Under tremendous pressure from President Trump, Mexico began to reactivate maquiladora activity on the border with the United States, especially in the automotive sector, to mitigate the effects of the economic paralysis but endangering the Mexican workers in those factors.

The pandemic has produced devastating effects in North America and has exposed the economic, social, and political unsustainability of the neoliberal paradigm.

Beyond the positive and negative results of the USMCA we will never know for sure the actual number of workers who were sacrificed to support the North American economy against COVID-19. The long-term impact of Trump's bullying tactics on his Canadian and Mexican neighbours will also be difficult to measure. Whether the election of Joe Biden will bring more friendly relations is also unclear since protectionist policies are likely to continue with his new administration. In this context Mexico and Canada will be ever-more reliant on building stronger relationships between themselves in the context of this hostile continental environment.

NOTES

1 The author appreciates the technical support of Carlos Julio Cantón Hernández.
2 In 1993, the Congress of the United States authorized a brief renewal of the Trade Promotion Authority (TPA) to conclude the negotiation of the Uruguay Round (1986–94). However, for Clinton's second term (1997–2001), Congress rejected this fast-track authority and halted pending negotiations with Chile and the Free Trade Area of the Americas.
3 Trade and Cooperation Arrangements were signed with MERCOSUR (1998) and with the Andean Community (1999); a Memorandum of Understanding on Trade and Investment with the Central America Countries (1998), the Canada-Chile Free Trade Agreement (1997); the Double Taxation Agreement with Brazil, Guyana and Mexico; the Free Trade Agreement Canada-Costa Rica (2002).

4 In 1994, the trade balances of the three countries added up to a total deficit of $1,815 million. Mexico had the largest deficit ($1,538.6 million) due to the crisis experienced in the country, which was extended to business partners. The Mexican deficit resulted, among other things, from the devaluation of the peso caused by the lack of international reserves. During 1995–7 the NAFTA members showed surpluses in the trade balance because of the internal changes of each economy. In Mexico, for example, the system of free floating of the peso was established, which translated into a loss of the value of the peso against the dollar, which led to an increase in exports and a trade surplus but a loss of the purchasing power of the population in general (Sánchez León 2010, 37–8).

5 According to the Secretariat of Economy, in 2015 bilateral trade between Mexico and the United States exceeded US$532 billion and was higher than the trade between the United States and Japan, Germany and South Korea altogether. In this sense, the United States consolidated as Mexico's first trade partner with 64 per cent of its total trade and 80 per cent of Mexican exports. For more see Secretariat of Economy (2016).

6 In Mexico, before 2013, there were no specified taxes on mining activities. The tax regime for mining companies was extremely permissive. Apart from the federal taxes mentioned above, the companies only paid a duty in terms of the hectares to be conceded. They did not pay any tax regarding extraction, transformation, or minerals sales. Before the reform to the Mining Law of 2013, companies only paid a duty to exploit the concession. The taxes and payment of duties by which mining companies must abide are federal. States and municipalities cannot levy taxes or mining-related activity duties.

7 Since 2019, the Mexican president repeatedly announced that the energy reform would be overturned, but no specific actions have been taken. If these did materialize, they would surely provoke numerous lawsuits by foreign companies against Mexico through the USMCA's dispute mechanism.

8 Among them, the elimination of NAFTA's Chapter 19 on dispute resolution in trade matters, removing tariff preferences, incorporates a chapter on anti-corruption, a chapter on energy and energy market access facilitation, and the elimination of Chapter 11 on Investor-State Dispute Settlement.

9 Clause 32.10 provides that entering into a free trade agreement by any of the Parties with non-market economies will allow the other Parties to terminate the agreement and replace it with a bilateral one. In practical terms, this clause broadly speaking refers to China, Cuba, and perhaps North Korea, considering these countries already belong to WTO. Mexico already has a

Bilateral Foreign Investment Protection Agreement with the People's Republic of China. The definition of a non-free-market country is set forth in accordance with the antidumping laws (price discrimination) of a Party, and if upon the execution of USMCA a Party does not have an FTA with that country.

10 Chrystia Freeland stated that the elimination of NAFTA's Chapter 11 had been one of her priorities during the renegotiation since Canada had lost the most after having faced the largest number of lawsuits in comparison to Mexico and the United States, for over US$300 million. Freeland asserted that she had received over 35,000 letters from Canadian citizens asking her to abandon investment arbitration.

REFERENCES

AMEXHI. n.d. "Industria de petróleo y gas natural de América del Norte: seguimiento a perspectivas sobre el TLCAN." Accessed 29 October 2021. http://www.amexhi.org/wp-content/uploads/2017/11/API-AMEXHI-CAPP-2nd-Position-Paper-on-NAFTA-20171115.pdf.

Arteaga, Alejandra. 2017. "Esto quieren los empresarios de EU en la renegociación del TLCAN." *Huffington Post* (25 April 25). Accessed 29 October 202. https://www.huffingtonpost.com.mx/2017/04/25/esto-quieren-los-empresarios-de-eu-en-la-renegociacion-del-tlcan_a_22055023/.

Ávila Pedro. 'Relación México-Canadá." Accessed 29 October 2021. http://www.connectionmexicoglobal.com/wp-content/uploads/2017/12/Relaci%C3%B3n-M%C3%A9xico-Canad%C3%A1.pdf.

Basave, Jorge, and María Teresa Gutiérrez Haces. 2017. "The Uneven Trends of Mexican MNEs: Between Sluggishness and Strength in the International Markets." Mexican Report on Mexican Multinationals, in Emerging Markets and Global Players Project, Columbia University, New York.

– 2018. "Uncertain Expectations of Mexican-American Economic and Trade Relations and Slowdown of Overall Mexican FDI." Mexican Report on Mexican Multinationals, in Emerging Markets and Global Players Project, Columbia University, New York.

Cámara de Comercio Servicios y Turismo Ciudad de México. "Si EU deja el TLCAN, Canadá se queda con México." Accessed 1 November 2021. https://www.ccmexico.com.mx/blog/noticias/si-eu-deja-el-tlcan-canada-se-queda-con-mexico.html.

Consulate General of Mexico in Montreal. "Programa de Trabajadores Agrícolas Temporales." Secretaría de Relaciones Exteriores. Accesssed 29 October 2021. https://consulmex.sre.gob.mx/montreal/index.php/es/ptat.

Cruz Serrano, Noé. 2017. "US-Canada and Mexico Commit to Energy Trade Alliance." El Universal (8 December). Accessed 29 October 2021. http://

www.eluniversal.com.mx/english/mexico-canada-and-us-commit-energy
-trade-alliance.

Cumbre de Negocios. n.d. "La relación México- Canadá: ¿nuevas
oportunidades ante la renegociación del TLCAN?" Accessed 1 November
2021. https://www.cumbredenegocios.com.mx/newsletter5/3.html.

El Diario. 2018. "Pide Texas mantener el capítulo 11 del TLC." (4 April).
Accessed 1 November 2021. http://diario.mx/Economia/2018-04
-04_41a5ff42/pide-texas-mantener-capitulo-11-en-tlc/.

Esquivel, Lindsay. 2018a. "Empresarios crean liga en favor del TLCAN;
Cámaras de México, Eu y Canadá." Excélsior (24 January). Accessed
29 October 2021. http://www.excelsior.com.mx/nacional/2018/01/24
/1215619.

– 2018b. "Sugieren a México salir del TLCAN: sindicato de la IP canadiense."
Excélsior (3 March). Accessed 29 October 2021. http://www.excelsior.com
.mx/nacional/2018/03/03/1223935.

Fariza, Ignacio. 2017. "La brecha salarial separa a México y Canadá en la
renegociación del TLC." El País México (24 September). Accessed 29 October
2021. https://elpais.com/internacional/2017/09/04/mexico/1504491392
_579796.html.

Fuentes, Monica. 2018. "México y Canadá rechazan modificar paneles de
arbitraje de TLCAN." Noticieros Televisa (29 January) Accessed
1 November 2021. https://noticieros.televisa.com/ultimas-noticias
/mexico-y-canada-rechazan-modificar-paneles-arbitraje-tlcan-wsj/.

Gabriel, Christina, and Laura Macdonald Laura. 2011. "Citizenship at the
Margins: The Canadian Seasonal Agricultural Worker Program and Civil
Society Advocacy." Politics & Policy 39.

Government of Canada. 2016. "Canada-Mexico Partnership- 2016, Annual
Report." Accessed 29 October 2021. http://www.canadainternational.gc.ca
/mexico-mexique/2016cmp-pcm.aspx?lang=eng.

Gutiérrez Haces, María Teresa. 2014. "NAFTA and Mining in Mexico: The
Emergence of New Actors and Conflicts." In L'ALÉNA á 20 Ans: Un Accord
en Sursis, un Modéle en Essor, Éditions IEIM, edited by Brunelle Dorval,
247–89. Montreal : l'Institut d'études internationales de Montréal.

– 2016a. "The Growing Presence of Canadian Mining Companies in Mexico
and the Dominance of Mexican Business Groups." In Latin American
Policy 7, no. 2.

– 2016b. Los vecinos del vecino. La Continentalización de México y Canadá en
América del Norte. Mexico City: ARIEL/PLANETA and Universidad
Nacional Autónoma de México.

Inversión Directa de Canadá a México. Accessed 1 November 2021. https://
www.gob.mx/se/acciones-y-programas/competitividad-y-normatividad
-inversion-extranjera-directa?state=published.

La Silla Rota. 2017. "México y Canadá mantienen posición sobre reglas de origen." (20 November). Accessed 1 November 2021. https://lasillarota .com/dinero/mexico-canada-mantienen-posicion-reglas-origen-tlcan -mexico-canada-reglas-origen/189581.

Lenard, Patti Tamara, and Christine Straehle, eds. 2012. *Legislated Inequality: Temporary Labour Migration in Canada.* Montreal-Kingston: McGill-Queen's University Press.

Mandel-Campbell, Andrea. 2008. "Why Mexicans Don't Drink Molson." Toronto: Douglas and McIntyre.

Morales, Roberto. 2016. "Canadá, protagonista en IED minera de México." *El Economista México* (4 November). Accessed 1 November 2021. https://www .eleconomista.com.mx/empresas/Canada-protagonista-en-IED-minera-de -Mexico-20161104-0121.html.

– 2017. "Canadá respalda a México en capítulo energético." *El Economista México* (6 December). Accessed 1 November 2021. https://www .eleconomista.com.mx/empresas/Canada-respalda-a-Mexico-en-capitulo -energetico-20171206-0021.html.

– 2020. "Canadá y México, limitados ante aranceles de EU." *El Economista México* (9 August). Accessed 1 November 2021. https://www.eleconomista .com.mx/empresas/Canada-y-Mexico-limitados-ante-aranceles-de-EU -20200809-0091.html.

Natural Resources Canada. 2017. "Memorándum de entendimiento entre el ministerio de recursos naturales de Canadá y la Secretaría de Economía de los Estados Unidos Mexicanos." (7 February). Accessed 1 November 2021. https://www.nrcan.gc.ca/node/19341/.

Notimex Canadá. 2017. "México y Canadá abogan por renegociación trilateral del TLCAN." (21 February). Accessed 1 November 2021. http:// revistaenterate.mx/index.php/dependencias-home/8442-mexico-y-canada -abogan-por-renegociacion-trilateral-del-tlcan.

Pablo, Sara. 2017. "México y Canadá apoyarían acuerdo bilateral si fracasa renegociación del TLCAN." VOANOTICIAS (24 October). Accessed 1 November 2021. https://www.voanoticias.com/a/tlcan-mexico-eeuu -canada-acuerdo-de-libre-comercio-donald-trump/4083677.html.

Panorama. 2017. "EEUU, México y Canadá retoman renegociación del polémico TLCAN." (11 October). Accessed 1 November 2021. http://www .panorama.com.ve/mundo/EE-UU-Mexico-y-Canada-retoman -renegociacion-del-polemico-TLCAN-20171011–0023.html.

Patiño, Dainzú. 2017. "La IP cabildea con empresarios de EU y Canadá a favor del TLCAN." Expansión (23 October 23). Accessed 1 November 2021. https://expansion.mx/economia/2017/10/23/ip-cabildea-con-empresarios -de-eu-y-canada-a-favor-del-tlcan.

Riquelme, Rodrigo. 2017. "Qué dicen los medios de EU y Canadá sobre la renegociación del TLCAN." *El Economista México* (22 August). Accessed 1 November 2021.https://www.eleconomista.com.mx/empresas /Que-dicen-los-medios-de-EU-y-Canada-sobre-la-renegociacion-del -TLCAN-20170822–0121.html.

Saldaña, Ivette. 2017. "Empresarios canadienses piden a Peña defenderse el TLCAN." *El Universal* (17 January). Accessed 1 November 2021. http:// www.eluniversal.com.mx/articulo/cartera/economia/2017/01/17 /empresarios-canadienses-piden-pena-defender-tlcan.

Sánchez León, Sarahí. 2010. "Crecimiento Económico y Comercio Exterior de México en el marco del Tratado de Libre Comercio con América del Norte, TLCAN, 1994–2008." Master's thesis, Colegio de la Frontera Norte, Tijuana.

Secretaría de Relaciones Exteriores (SRE). 2016. "Hoja informativa México-Canadá. Una relación más amplia y profunda." Accessed 1 November 2021. https://www.gob.mx/sre/acciones-y-programas/mexico-y-canada -una-relacion-amplia-y-profunda-7990?state=published.

– "México y Canadá: 72 Años de Una Relación Estratégica," Accessed 1 November 2021. https://www.gob.mx/cms/uploads/attachment/file /86322/Hoja_informativa_Mexico_y_Canada_esp.pdf.

– n.d. "México y Canadá: Una Relación Amplia y Profunda." Accessed 1 November 2021. https://www.gob.mx/sre/acciones-y-programas /mexico-y-canada-una-relacion-amplia-y-profunda-7990?state= published.

Secretariat of Economy. 2015. "Inversión Extranjera Directa." (10 June). Accessed 1 November 2021. https://www.gob.mx/se/acciones-y -programas/competitividad-y-normatividad-inversion-extranjera -directa?state=published.

– 2016. "Relación México - Estados Unidos." (21 July). Access 1 November 2021. https://www.gob.mx/epn/articulos/relacion-mexico-estados-unidos -49795.

– 2017. "Inversión Extranjera Directa en México y en el Mundo." (8 December). Accessed 1 November 2021. https://www.gob.mx/cms /uploads/attachment/file/279079/Carpeta_IED.pdf.

– 2018. "Inversión Directa de Canadá hacia México." Accessed 1 November 2021. https://www.gob.mx/cms/uploads/attachment/file/302801 /Distribucion_de_inversion_de_Canad_.pdf.

Secretariat of Energy. n.d. "Se fortalece la cooperación México- Canadá en energías renovables." Accessed 1 November 2021. https://www.gob.mx /sener/prensa/se-fortalece-la-cooperacion-mexico-canada-en-energias -renovables.

Segal, Stephanie. 2018. "USMCA Currency Provisions Set a New Precedent." Center for Strategic and International Studies (5 October). Accessed 1 November 2021. https://www.csis.org/analysis/usmca-currency -provisions-set-new-precedent.

Servicio Geológico Mexicano Anuario Estadístico de la Minería Mexicana. 2016. "Anuario Estadístico de la Minería Mexicana, 2016." Accessed 1 November 2021. https://www.sgm.gob.mx/productos/pdf/Anuario _2016_Edicion_2017.pdf.

Sinclair, Scott. 2018. "USMCA Strikes a Welcome Blow against Investor-State Dispute Settlement." The Monitor (10 October). Accessed 1 November 2021. https://monitormag.ca/articles/usmca-strikes-a-welcome-blow -against-investor-state-dispute-settlement.

Tele Sur. 2017. "Canadá deja solo a México en renegociación." (25 January). Accessed 1 November 2021. https://www.telesurtv.net/news/Canada-deja -solo-a-Mexico-en-renegociacion-comercial-del-TLC-20170125–0008.html.

20 Minutos. 2017. "Es minería un tema especial en relación México-Canadá: ProMéxico." (6 March). Accessed 1 November 2021. https://www .20minutos.com.mx/noticia/195148/0/es-mineria-un-tema-especial-en -relacion-mexico-canada-promexico/.

Vanegas García, Rosa María. 2019. "Cuatro décadas del Programa de Trabajadores Agrícolas Temporales México-Canadá: 1974–2014." Mexico City: Secretaría de Cultura de México y el Instituto Nacional de Antropología e Historia.

Vanguardia MX. 2017. "Empresarios mexicanos piden que EU y Canadá, no se metan en tema laboral." (3 September). Accessed 1 November 2021. https://www.vanguardia.com.mx/articulo/empresarios-mexicanos-piden -que-eu-y-canada-no-se-metan-en-tema-laboral.

Winfield, David. n.d. "Relaciones bilaterales Canadá–México." Accessed 1 November 2021. https://revistadigital.sre.gob.mx/images/stories /numeros/n38/winfield.pdf.

Wood, Duncan. 2012. "Canada-Mexico Relations Moving Beyond 65 Years of Stunted Growth." In *Canada Looks South: In search of an Americas Policy*, edited by Peter McKenna, 117–48. Toronto: University of Toronto Press.

9 "Wrapped in the Canadian Flag": Precious Metals Mining and Canada's Deadly Diplomacy in Latin America

JEN MOORE

Much of the discussion in Canada concerning Canadian mining companies abroad has tended to focus on the abuses committed by certain corporations and the lack of access to justice and remedy in Canada – or just about anywhere – for those harmed. Less attention has been directed to the role Canadian authorities have played to facilitate expansion of a mining model in Latin America that has been the focal point of intensifying criminalization and violence in the region. The "mining model" refers to industrial mining as a highly intensive extractive activity that takes place within a legal, economic, and political framework oriented to serve global commodities markets and maximize mining companies' pursuit of extraordinary profits. At the same time, it externalizes serious social and environmental impacts for the long-term onto affected communities and, disproportionately, Indigenous peoples.

Canadian state support for this model has included going to bat for individual companies mired in conflict, as well as advocating for legal and regulatory frameworks that enable private investment in mining sectors around the world. This is what the Canadian government calls "economic diplomacy." It is defined in the 2013 Global Markets Action Plan, which states that "all diplomatic assets of the Government of Canada will be marshaled on behalf of the private sector in order to achieve the stated objectives within key foreign markets" (DFATD 2013, 11). Given the relative importance of investment in mining as a large component of Canada's foreign direct investment abroad, this adds up to more support for Canadian mining companies, despite the proliferation of conflict over the mining model in Latin America and elsewhere.

Kelvin Dushnisky, the former President of Barrick Gold – a company that has been at the centre of social and environmental conflicts in countries such as the Dominican Republic, Peru, Argentina, and Chile – recently commented to the *Northern Miner* on the importance of

Canadian government support to companies like his: "There's a strong element of support from the Canadian government, and in our ability to wrap ourselves in the Canadian flag. We found that to be extremely effective as we operate abroad" (Cumming 2017; see also Barrick 2018). Dushnisky affirms what experience in the region and investigations based on access to information requests have been revealing about how Canada's whole-of-government approach in Latin America is designed to promote laws favourable to mining companies and to troubleshoot on behalf of those who are in conflict with Indigenous peoples, affected communities, and workers.

Additionally, Canadian authorities have been steadfast proponents of investor protection agreements that allow companies to sue governments for hundreds of millions of dollars when states adopt measures to protect people or the environment, regardless of whether companies have followed local laws and regulations. Dushnisky indicated that this is another important element of Barrick's risk management strategy when determining where to invest: "We also look to jurisdictions where investment treaties could exist, ranging from Canada, which uses foreign investor protection agreements or even better, where we have free trade agreements with strong investor protection chapters. Those help us" (Cumming 2017). Such legally enforceable rights for corporations deepen asymmetries between mining companies and communities who are fighting to protect areas from mining and who most often lack effective means to obtain justice for mining harms.

In the following section, the way Canadian mining investment has dramatically expanded in Latin America over the last couple of decades in parallel with intensifying criminalization, violence, and militarization in affected areas is described in general terms. Following that, various examples of Canadian "economic diplomacy" to promote Canadian mining interests are explored, first at the policy level and then at the project level. Finally, arguments that Canadian authorities frequently use to justify state support for the mining model are discussed in juxtaposition with the reality that communities are struggling with on the ground where the Canadian state has intervened in favour of Canadian domiciled or financed corporations.

Growing Canadian Investment and Intensifying Mining Violence in Latin America

The advantages the Canadian state affords to the globalized mining industry has given Canadian companies a frontline role in a deepening social, ecological, and political crisis in many countries in Latin America.

Domestically, Canadian stock exchanges have specialized in financing the early stages of mining. As a result, approximately 60 per cent of the world's mining companies, or over 1,300 companies, were listed on Canadian stock exchanges by the end of 2015. The vast majority of these firms – roughly 90 per cent – do not have any mining projects in operation. Rather, they are prospecting and exploration firms that speculate on potential mineral finds. This helps explain why Canadian companies are so often the first that mining-affected communities encounter, well before any mine is built, but often at the moment that serious conflicts begin to arise.

Internationally, Canada's foreign-policy orientation for the sector has been to urge host countries towards greater extractive industry dependence through privatization of mineral extraction and export-driven and commodity-fuelled economic growth. As resistance and conflict have multiplied in response to mining expansion in Latin America since the 1990s, this has given rise to intensified processes of demonization, criminalization, and militarization against communities, social movements, and other sectors of society that question this development model (MiningWatch Canada and ICLMG 2015). Webber and Gordon (2016, 28) describe this process as "the imposition of a logic of accumulation by dispossession, pollution of the environment, reassertion of power in the region by multinational capital, and new forms of dependency. It also, necessarily and systematically involved what we call militarized neoliberalism: violence, fraud, corruption, and authoritarian practices on the part of militaries and security forces. In Latin America, this has involved murder, death threats, assaults and arbitrary detention against opponents of resource extraction."

The Canadian government does not publicly track the incidents of conflict and violence around Canadian mining projects in the region. However, its prevalence is reflected in a report from the Justice and Corporate Accountability Project at Osgoode Hall Law School (Imai, Gardner, and Weinberger 2016). Based on thousands of hours of research by a group of volunteer law students from five Canadian universities, and providing what is inevitably understated and preliminary data, *The Canada Brand* finds that between 2000 and 2015 at least 44 people were killed, 403 injured, and 709 criminalized in 13 countries in Latin America, in events related to projects belonging to 28 Canadian mining companies. Notably, this study did not include incidents of violence where people received death threats, had their crops or property destroyed, were forcibly displaced, suffered murder attempts without physical injury, or became sick or suffered psychological trauma. Cases were included from Mexico, all Central American countries except for Costa Rica, seven South American countries, and the Dominican Republic.

Acknowledging that the numbers presented in *The Canada Brand* underestimate the gravity and intensification of violence occurring at Canadian mine sites in Latin America, the authors reference escalating violence, extortion, and forced displacement where Canadian mining investments are expanding in areas where organized crime has exerted territorial control. *The Canada Brand* provides the example of the Los Filos, a gold mine in Guerrero, Mexico, which was owned and operated by Goldcorp until 2017 when it was sold to Canadian company Leagold Mining Corp. (Leagold was subsequently acquired by Equinox Gold in late 2019.) The authors note that up to 2015 there were at least seventeen deaths – nearly 40 per cent of the number of deaths recorded in the entire report – as well as three disappearances, eight injuries, and hundreds of families forcibly displaced. The authors state that this data is not included in the report's overall statistics "because the situation is very complex and involves the presence of armed gangs and corrupt state armed forces" (34). Nonetheless, as the authors observe, the rise in violence has paralleled the operations of the Los Filos mine: "The mine is located between the towns of Mezcala and Carrizalillo, which have reportedly seen an increase in the presence of organized crime since the company began commercial production in 2008. Reports indicate that criminal groups have controlled and terrorized local communities, especially the town of Carrizalillo. They have used fear to extort local mine workers and those receiving payments from the company for the use of their land" (33).

Nonetheless, while reporting tens of millions of dollars in annual earnings from the Los Filos mine, Goldcorp asserts that it "'[works] for mutual benefit in politically stable jurisdictions' and that it is committed to responsible mining" (quoted by Imai, Gardner, and Weinberger 2016, 33). Furthermore, the company cynically reports being a signatory to the World Gold Council's Conflict-Free Gold Standard (n.d.), which it says it applied at the Los Filos mine during the height of violence and forced displacement in 2015, at which time three of its workers were kidnapped and murdered and half of the community of Carrizalillo fled their homes (Goldcorp 2016). This same year, community members complained that the company failed to respond to their requests for assistance, to which the company stated that it does not bear responsibility for violence occurring outside its mine gates (Imai, Gardner, and Weinberger 2016). The question here, however, is not so much whether Goldcorp is directly responsible for these murders, disappearances, and forcible displacement, but rather with what sense of entitlement – and absolute callousness – it is profiting richly when a virtual war zone has arisen outside its mine gates since going into operation. With what notion of responsibility does it freely operate – with armed security and government support – in an area where the affected communities are terrorized without the opportunity to fully exercise their self-determination or pressure government authorities to address mining harms, including out-of-control violence, particularly when these same authorities are believed to be acting in collusion with organized crime? Furthermore, in a context in which families and workers are being extorted and their lives put in constant jeopardy, how is it possible that the mining company and its activities are thriving without some direct or indirect agreement or arrangement with the same criminal groups or corrupt state armed forces? As explored later in this article, the creation of such a state of terror has not impeded Canadian embassy officials from lending their support to Canadian mining companies.

In parallel to skyrocketing risks for affected communities, more than twenty years of Canadian promotion of the globalized mining sector has given rise to spectacular growth in Canadian mining and banking investment in Latin America with tremendous benefits for transnational companies such as Goldcorp, Barrick Gold, and Yamana Gold (Gordon and Webber 2016). From 1990 to 2013, Gordon and Webber (2016) find that Canada's overall Foreign Direct Investment in Latin America grew by 2,000 per cent, while US investments increased 555 per cent during the same period.[1] They further calculate for 2012 that Barrick, Yamana Gold, and Goldcorp earned a combined $2.8 billion in net profit from their operating mines, and Canadian mining companies worldwide earned $19.3 billion. Between 1998 and 2013, they determine that these three companies averaged a 45 per cent rate of profit – which they call "super-profits" – on their operating mines when the Canadian economy's average rate of profit was 11.8 per cent. By comparison, the Canadian government sent $187.7 million in overseas development aid to Latin America in 2012 (it is not clear how much of this was destined for programmes intended to support the Canadian mining sector through mining-related training, infrastructure, and reform projects). Meanwhile, Webber and Gordon observe that remittances from migrants in Canada to Latin America this same year, were $798 million.

The spectacular costs, however, to Indigenous peoples and affected communities have been mounting and are rarely addressed, nor are the macroeconomic risks that countries are exposed to as a result of increased dependency on mineral extraction, commonly understood as the "resource curse." The resource curse usually refers to the macroeconomic pitfalls of reliance on non-renewable resource extraction or the way that such dependency tends to undermine economic development on a national scale. The curse makes national economies vulnerable to the boom and bust cycles of mineral prices on the global market. This include the overevaluation of local currency when mineral prices are high, which can negatively impact other sectors of a national economy, known as the "Dutch Disease." Dependency on non-renewable resources also tends to foster corruption and misspending during boom periods and overall diminished attention to more sustainable economic sectors. Anthropologist Stuart Kirsch (2014, 33) argues that there is also a local side to such dependency, which he describes as "the microeconomics of the resource curse," referring to the incomplete accounting of the social and environmental costs of mining projects at the local level "that might shift their value from the positive to the negative side of the ledger." Garibay and Balzaretti use the concept of "negative reciprocity" or "the attempt to obtain something with impunity in exchange for nothing" to describe this same phenomenon (2009, 92).

This is the dangerous and unjust development model the Canadian state has contributed to entrenching and defending in Latin America and other parts of the world.

Canadian Economic Diplomacy to Entrench the Mining Model at the Policy Level

Since the 1980s, the World Bank Group, with the support of a number of governments in the Global North, including Canada, has been driving efforts to institute reforms in the mining codes of country after country to facilitate private foreign investment (CNCA 2007; see also Moody 2007). By the early 2000s such mining reforms had been adopted in an estimated 100 countries around the world. These reforms removed restrictions on foreign ownership and repatriation of profits, lowered rates of taxation and royalties, ended local sourcing and hiring obligations, eroded labour protections, and streamlined administrative processes to make permitting easier (Szablowski 2007). The Canadian government provided strategic backing to the formation of key policies and projects in a number of countries using a whole-of-government approach, employing aid, diplomacy, and trade relations to influence policy change (Gordon and Webber 2016).

For example, in Colombia, Canadian aid paid for a technical assistance project to assist with mining law reforms through which intermediaries or agents for Canadian companies were contracted as experts (MiningWatch Canada, CENSAT-Agua Viva, and Inter Pares 2009). These reforms were approved in 2001, in the context of a war in which rural, Indigenous, and Afro-Colombian peoples are disproportionately represented among the millions of victims of forced displacement and systematic violence (Americas Policy Group 2012). Notably, of the some 3–5 million internally displaced people in Colombia, the great majority[2] have been forced out of mineral and hydrocarbon rich areas (Vicente et al. 2011). Developed with the support of lawyers who work closely with the mining industry, the reforms paved the way to dismantle the state mining company, criminalize the artisanal and small-scale mining sector, subordinate local land-use plans to mining interests, and limit the role of environmental authorities (Americas Policy Group 2012). Under the administration of President Álvaro Uribe, during a period of high commodity prices, they also paved the way for rapid territorial expansion of mineral and oil concessions which rose to an estimated 40 per cent of national territory (Vicente et al. 2011, 8). Meanwhile, Canadian-financed companies have been well positioned to benefit (MiningWatch Canada, CENSAT-Agua Viva, and Inter Pares 2009).

In post-military coup Honduras, Canadian authorities used aid dollars and diplomatic influence to prevent community and civil society–driven reforms from being introduced to the mining code, and to develop a law more favourable to industry.

Hondurans realized shortly after Goldcorp's San Martín mine in the Siria Valley went into production around 2000 that their mining law, rushed through in the wake of Hurricane Mitch in 1998, provided no recourse for communities suffering negative impacts. As a result, they pressed for reforms to ban open-pit mining and the use of certain toxins in mine processing and called for communities to have a decisive say early in the process that would determine whether mining could take place on their lands. They made headway. In 2004 their government put a moratorium on new mining projects, and by 2009 a draft law incorporating their demands was ready for debate. But this came to a halt with the June 2009 coup against President Mel Zelaya and the systematic human rights violations against the coup resistance that followed, illegal processes which Canada failed to denounce and sanction. The mining bill was never debated.

Rather, after the November 2009 elections – widely criticized as illegitimate – Canadian authorities wasted no time in beginning to lobby for a mining law to suit the industry and for the lifting of the moratorium (MiningWatch Canada 2012). Gordon and Webber's *Blood of Extraction* (2016), for which the authors undertook a detailed review of documents obtained from access to information requests, provides helpful insights into how Canadian authorities took advantage of the political opportunities afforded by the coup to push forward measures that favoured big business.

A key goal for the Canadian government, according to one embassy memo, was "[to facilitate] private sector discussions with the new government in order to promote a comprehensive mining code to give clarity and certainty to our investments" (68). Another embassy record said that mining executives were happy to assist with the writing of a new mining law that would be "comparable to what is working in other jurisdictions" and was to be developed with a resource person with whom their "ideologies aligned" (ibid.).

In a highly authoritarian and repressive context, and under the banner of corporate social responsibility, the Canadian Embassy – with additional support from Canadian ministerial visits, a Honduran delegation to the annual meeting of the Prospectors & Developers Association of Canada (PDAC), and overseas development aid for technical support – managed to get the desired law passed in early 2013, lifting the moratorium. The new law opens the gates to new mining projects in what has become one of the most violent countries in the region and is a major setback compared to what Hondurans had been proposing (MiningWatch Canada 2013a). The law opens the door to open-pit mining, many community water sources are not protected, mining is not prohibited in populated areas, leaving them vulnerable to the risks of land expropriation and displacement, and community rights to decide whether mining can happen are narrowly defined (ibid.). Despite high levels of corruption among police and increasing military cooperation with police, the mining law establishes a 2 per cent security tax on mining activities that will provide a direct incentive to security forces to protect corporate interests. In 2014 two civil society organizations challenged the constitutionality of the mining law with the Constitutional Court (MiningWatch Canada 2015a). In 2017, their arguments were partially accepted by the court, including challenges related to the allowance for open-pit mining, reduced corporate taxes, and the lack of prior consultation with affected communities (Contracorriente 2017).

The Canadian government did not stop with the mining law. In June 2014, with full support from Liberals and Conservatives in the House of Commons and the Senate, Canada ratified a free trade agreement with Honduras, effectively declaring that "Honduras, despite its political problems, is a legitimate destination for foreign capital," comment Gordon and Webber (2016, 80). The agreement helps to lock into place the mining reforms, providing Canadian mining investors with recourse to binding international arbitration that they can use to threaten or sue the Honduran state should it ever take measures that could impede their investments in one form or another.

The possibility that this could happen is not far-fetched. Such provisions in investor protection agreements are increasingly used by oil, gas, and mining companies to sue states for outlandish amounts of money when they make decisions the companies do not like (CNCA 2007). For example, Pacific Rim Mining sued the state of El Salvador for US$301 million for not having granted it a permit to put a gold mine into operation, even though the company did not meet the regulatory requirements to obtain the permit (Moore et al. 2014). While the company did not win the case, it cost the state millions of dollars to fight the suit over the course of seven years, during which time a chill was put on public policymaking in the area of mining. During this period, four community members who were part of the resistance to the company's project were murdered and many others threatened. As of March 2013, there were 169 cases pending at the most frequently used tribunal, the International Centre for Settlement of Investment Disputes (ICSID), of which 60 (35.7 per cent) were related to oil, mining, or gas (Anderson and Perez-Rocha 2013). By comparison, in 2000, there were only three pending ICSID cases related to oil, mining, or gas (ibid.).

In contrast to how Canada has strongly aligned itself with Latin American regimes openly supportive of militarized neoliberalism, such as Colombia and Honduras, the experience in Ecuador under the administration of President Rafael Correa illustrates how Canada considers "any government that does not conform to the norms of neoliberal policy, and which stretches, however modestly, the narrow structures of liberal democracy ... a threat to democracy as such" (Gordon and Webber 2016, 212).

Canada played a key role in limiting modest reforms advanced under the administration of President Correa, thereby undermining the opposition of affected communities and social movements to opening the country to large-scale mining. A critical moment in this process occurred in mid-2008 when a constitutional-level decree was issued in response to local and national mobilizations against mining, at a time when Canadian-domiciled companies made up a vast number of mining firms in the country. The Mining Mandate, as it was called, would have extinguished most or all of the mining concessions that had been granted in the country without prior consultation with affected communities or that overlapped with water supplies or protected areas; and it included additional criteria for revocation (Moore and Velásquez 2012). It also set in place a short timeline for the development of a new mining law.

The Canadian embassy immediately went to work. Meetings between Canadian industry and Ecuadorian officials, including the president, were set up to ensure a privileged seat at talks on the new mining law (ibid.). Gordon and Webber's review of documents obtained under access to information requests further reveals that the embassy even helped organize pro-mining demonstrations together with industry and the Ecuadorian government. Embassy records describe their intention "to create sympathy and support from the people" as part of "a pro-image campaign," which included "an aggressive advertisement campaign, in favour of the development of mining in Ecuador" (Gordon and Webber 2016, 231). Meanwhile, behind closed doors, industry threatened to bring international arbitration against Ecuador under a Canada–Ecuador investor protection agreement, which a couple of investors eventually did (MiningWatch Canada and ICLMG 2015).

Ultimately, Gordon and Webber (2016, 233) conclude, Canadian diplomacy "played no small part" in ensuring that the Mining Mandate was never applied to most Canadian-owned projects and that a relatively acceptable new mining law was passed in early 2009. While embassy documents show the Canadian government considered the law useful enough to "open the sector to commercial mining," it was still not business-friendly enough, particularly because of the higher rents the state hoped to reap from the sector. As a result, the embassy kept up the pressure, including using the threat of withholding badly needed funds for infrastructure projects until mining company concerns were addressed and dialogue opened with all Canadian firms.

The pressure from Canadian industry continued, eventually achieving its desired reforms in 2013 that weakened environmental requirements and the tax and royalty regime in Ecuador (MiningWatch Canada and ICLMG 2015). Meanwhile, as the door was pried open to the mining industry, mining-affected communities and supporting organizations were feeling the walls of political and social organizing space cave in as they faced persistent legal persecution and demonization from the state itself, while the negative impacts of the country's first open-pit copper mine began to be felt (ibid.).

Canadian Economic Diplomacy to Address Mining Conflict at the Local Level

In addition to lobbying efforts on behalf of the Canadian mining industry at the policy level, the Canadian government also provides a myriad of services to mining companies, including help to troubleshoot when they face conflicts with Indigenous peoples, affected communities, and workers at the project level. MiningWatch has identified over a dozen such examples (MiningWatch Canada 2013b), but few have been documented in as much detail as the cases of the Canadian Embassy in Mexico during conflicts surrounding Blackfire Exploration's barite mine in Chiapas and Excellon Resources' silver mine in Durango. As lawyer Charis Kamphuis has observed, these cases and others yet unpublished, demonstrate how Canadian embassies fail to consider respect for human rights when determining company eligibility to receive, or continue to receive, their services (Kamphuis 2018).

Notably, as of 2018, 155 of 233 foreign mining companies in Mexico have offices in Canada, that is, over 65 per cent (Secretaría de Economía México 2018). Canada is the largest foreign investor in Mexico's mining sector (Dawson 2014), and Canada has greater mining assets in Mexico than in any other country in the Americas, outside of Canada (Natural Resources Canada 2018).

The case of Blackfire Exploration in Chiapas, Mexico, illustrates how embassy support not only fails to assist land and environment defenders but might actually contribute to endangering their lives. Blackfire Exploration operated a barite mine in the municipality of Chicomuselo, Chiapas, from late 2007 until late 2009 when the mine was shut down for environmental reasons mere days after the shooting murder of Mariano Abarca on 26 November 2009.

Mariano Abarca, a father of four and a restaurant owner in the town of Chicomuselo, was an important community leader in the opposition to Blackfire's Payback mine. In July 2009 he participated in a delegation that travelled from Chicomuselo to Mexico City to protest in front of the Canadian Embassy. There he was videotaped speaking to an embassy representative when he stated that the company had broken its promises to provide work to everyone in the Ejido Grecia; that infrastructure in Chicomuselo had been damaged by the company's trucks; and that the community was highly concerned about environmental contamination given the importance of nearby rivers that flow from the Sierra Madre highlands of Chiapas (Rivera Sierra 2009).

On film, Abarca further alleged that Blackfire workers were acting as "thugs" against protesters. He concluded by stressing that community members who spoke out about problems were at personal risk: "Some of us in the movement have received threats and we don't think it's fair that foreigners come in creating conflict, while taking the wealth back to their country" (ibid.).

Three weeks later, undercover police detained Abarca in response to a complaint filed by Blackfire's Public Relations Officer, Luis Antonio Flores Villatoro. The complaint alleged that Abarca was responsible for crimes of illicit association, organized crime, attacks on communication routes, damages against the company and disturbing the peace, and threats against bodily integrity, as well as collective integrity and the integrity of state heritage (Procuraduria General de Justicia del Estado Mexico 2009). After being held for eight days, Abarca was released without charge for lack of evidence (United Steelworkers, Common Frontiers, and MiningWatch Canada 2010, 11). In a videotaped interview recorded at that time, Mariano said that if any harm should befall him, his family, or other activists, the community would blame Blackfire (Otros Mundos AC/Chiapas – Amigos de la Tierra México 2010).

On 27 November 2009, about four months after protesting in front of the Canadian Embassy, a male assailant shot Mariano Abarca in the back at close range in front of his house. The three individuals detained immediately following the murder all had connections with the company, although none of those named by the Abarca family and activists closely following the conflict were ever investigated (United Steelworkers, Common Frontiers & MiningWatch Canada 2013), except for one who was jailed in what is widely believed to be a case of political persecution (Mandujano 2012).

Based on documents later obtained from the Canadian Department of Foreign Affairs, the Canadian Embassy was aware of the tensions around Blackfire's mine (United Steelworkers, Common Frontiers, and MiningWatch Canada 2013). Even before the mine went into operation, the embassy knew the company had difficulties reaching agreements with local communities and exerted diplomatic pressure on Chiapas state officials to enable the mine to get up and running. Once in operation and until roughly mid-2009 the embassy monitored media reports about thousands-strong protests in Chiapas, received documents expressing opposition to the mine, made reports of months-long blockades, heard the testimony of Mariano Abarca in July 2009 about armed workers being used to intimidate peaceful protesters, and received some 1,400 letters after Abarca's 2009 detention that expressed dire worry about his wellbeing. These many pieces of verifiable information should have represented red flags for the embassy regarding Blackfire's operation. Instead, an embassy staff member dismissed them as nothing but tactics to "shake down" the company for more money (ibid.).

After Abarca's detention in August 2009, the embassy undertook to gather information and communicate with certain federal and Chiapas state agencies. Its approach, however, was oriented towards dispelling doubts over the legitimacy of Blackfire's operation and promoting the company's characterization of the protests. Although we do not have a full record of the embassy's meetings with state officials, in communications around embassy officials' fact-finding mission to Chiapas in October 2009, there is no evidence that they tried to speak with affected community groups and activists directly involved in the conflict. Instead, they raised concerns with the state government about possible increases in royalty payments levied on Blackfire and called on government officials to quell the protests (ibid.). It is conceivable that the embassy's support for Blackfire may have not only denied support for Mariano Abarca's safety, but even put his life at greater risk. This allegation is at the centre of a current complaint that was presented by the Abarca family in February 2018 to the Public Service Integrity Commissioner (MiningWatch Canada 2018a) and is now being considered by the Canadian Federal Court of Appeal (MiningWatch Canada 2019).

During the period from mid- to late 2009 when the Canadian embassy came to the defence of Blackfire Exploration, the same documentation obtained from the federal government shows that the embassy was also troubleshooting for several other firms. Within a period of just two weeks in early October 2009, as described in an email from the embassy's Trade Commissioner titled *Troubleshooting for Canadian Mining Companies*, the embassy intervened at senior levels for four Canadian mining investments in Mexico.[3] These included backing company interests in connection with: a year-long strike against the La Guitarra mine in the state of Mexico (First Majestic Silver Corp. n.d.) then owned by Genco Resources;[4] lack of prior community consent at Fortuna Silver's San José mine in Oaxaca; and an unresolved land-titling dispute at Oromex Silver's Tejamen project in Durango.[5]

A second well-documented conflict in Mexico also captured through an access to information request concerns landowners from the Ejido La Sierrita and workers from Local 309 of the National Miners Union at Excellon's La Platosa mine in Durango. The workers and landowners undertook a peaceful protest on private property outside the mine gates for several months in the summer of 2012, after filing two formal complaints in Canada alleging serious land and labour rights violations without result.

The subsequent 250-page release from the Canadian federal government revealed that, despite full knowledge of these complaints and Excellon's refusal to engage in dialogue to address them, the Canadian Embassy planned to share information with Excellon that was gathered from community members and their legal counsel without their consent, while helping the company forge high-level connections that led to violent repression against the protest. In contrast, no evidence was found of the Canadian embassy encouraging the company to act responsibly or to respect international standards, as it purports to do (Voices-Voix 2017; MiningWatch Canada and United Steelworkers 2015a).

Further, when the embassy was forewarned that Mexican police, army, and government officials were meeting to plan to evict the protest in response to the embassy and company lobbying, one trade commissioner wished the company well the night before forces moved in on the protest camp. Meanwhile, the embassy showed no regard for the safety of peaceful protestors, despite the frequency of injuries and killings of human rights activists, journalists, and community leaders in Mexico.

The disdain and repression that the Ejido La Sierrita members experienced in 2012 ended Excellon's welcome in their community. They have since taken action to rescind their contract with Excellon to try to end their relationship with the company (MiningWatch Canada and United Steelworkers 2015b).

Other efforts to document embassy–company relations have been less fruitful, with responses to access to information requests generating only highly redacted or partial documentation (see Weisbart 2018), and there are few indicators of any meaningful shift in embassy behaviour since the change in government in Canada in 2015. On the contrary, the Canadian embassy in Mexico has been willing to make public manifestations of support for expanding Canadian mining investment in the most dangerous parts of the country with no apparent regard for community members who have been risking their lives to speak out about labour rights violations, water contamination, and high levels of insecurity.

Notably, in April 2016, Canadian Ambassador to Mexico Pierre Alarie joined the Governor of Guerrero to celebrate the inauguration of Torex Gold's El Limón-Guajes gold mine in Cocula, Guerrero (Villagomez 2016) (the same Cocula where the Mexican government has alleged – against all scientific plausibility – that forty-three disappeared students from a rural teachers' college in Ayotzinapa were burned in a garbage dump). It did not seem to matter to the Canadian Embassy that a mine manager had already been murdered and workers kidnapped in connection with Torex Gold's mine, or that communities had been protesting over unfulfilled agreements, contaminated water, and health problems. Nor was it apparently of concern that, one month after the ribbon-cutting ceremony, the same communities complained in the press about being under siege from organized crime and the lack of a response from the governor to ensure their safety (Flores 2016), while at the same time the mine site was reportedly guarded by all levels of state armed forces.

Rather, in keeping with statements first made in 2015 (MiningWatch Canada 2015b), the Canadian Ambassador was willing to write off such matters as generalized problems of insecurity, suggesting to the Mexican press that a direct connection could not be made with Canadian mining companies (MiningWatch Canada 2016). But once again, as with the case of Los Filos mentioned earlier, it is hard to conceive that when companies are operating in an area known to be controlled by organized crime acting in collusion with state forces, where communities and workers are being extorted for what money they make from mining and other activities, and when extreme violence and massive displacement occur with frequency in the area, companies are not contributing to such horrors – whether directly or indirectly – at the same time that they are profiting from them. While such things are difficult to prove in a context of impunity that brings deadly consequences for those who make direct pronouncements, prior to the inauguration of the El Limón-Guajes mine, there was public speculation that Torex Gold had enjoyed business relationships with an organized crime group "La Familia" (Inka Kola News 2015). Did the Canadian Embassy not notice those reports before it decided to attend the inauguration of the new mine? Or did it not care? When this is the reality that local communities are facing and the embassy is still willing to step up on behalf of Canadian companies, there is clearly little stopping them from operating wherever they happen to find a mineral deposit, regardless of whether it is in a so-called politically stable jurisdiction and what such "stability" might entail.

The World Needs More Canada?

On 17 January 2018 then Minister of International Trade François-Philippe Champagne hosted a press conference to announce the creation of a human rights ombudsperson to investigate complaints from workers and communities facing abuses of Canadian corporations around the world, particularly extractive industry firms. The nature and limitations of this proposed office are beyond the scope of this paper, although it is important to point out that the same civil society coalition that lobbied for this office for over a decade is now warning mining-affected communities against its use (CNCA 2020). However, the Minister's statements put the government's beliefs about Canada's role in the globalized mining sector on prominent display.

Ignoring vast disparities between the super-profits of companies like Barrick Gold or Goldcorp and the negative balance that affected communities are left with, Minister Champagne opened his press conference asserting that the "Canadian brand" is an asset to companies and communities alike. "To fly the Maple Leaf means something, it means being associated with a set of values based on dignity, respect, diversity and tolerance. And it means a commitment to inclusive and sustainable economic growth where the benefits are shared by the many and not the few," he proudly stated (McSheffrey 2018). The Minister acknowledged that there were some problems for which reason the Ombudsperson was being announced, but the fundamental problem principally lay with other countries and weak legal frameworks in his view, not Canada and the economic agenda it was flogging in the world. "Canadian companies operate in many countries where there are great benefits to local economies and communities. However, in some of these countries, they still lack the legal and regulatory guarantees we know we all enjoy in Canada. Canada is at the forefront of efforts to strengthen the global rules-based order," he remarked (ibid.). He further expressed confidence that the industry would help seek solutions where needed, except perhaps for a few bad apples. "While I am confident that for the vast majority of companies [it] will be the case [that they will cooperate in good faith with the ombudsperson], I have heard the concerns that there could be at some point a company that refuses to fully engage and that could damage the strong Canadian brand we all gain from. This also concerns me because it is my job to promote the brand and to protect it" (ibid.). With this statement, the focus was deftly shifted to the problem that *some* Canadian mining companies were creating for Canada's reputation, rather than calling attention to those workers, Indigenous peoples, and affected communities seeking a modicum of justice for the multitude of mining harms and injustices they had faced.

In sum, the Minister painted the mining model as a win-win for companies and communities, in which most Canadian mining companies could be expected to espouse Canadian values abroad; in which the bigger issues with poor laws are for other governments to solve, although the Canadian government could be counted on to come out in support of stronger law and order; and where conflicts arose, companies could largely be expected to pitch in constructively.

Nonetheless, evidence from cases such as those involving Blackfire and Excellon demonstrate that Canadian embassies have Canadian corporate interests at heart when they keep a close eye on their mining operations and troubleshoot for them when conflicts arise, despite specific knowledge of community and worker complaints and related risks of repression or violence. Canadian representatives have more recently demonstrated in cases like that of Torex Gold in Guerrero, that the embassy will show up to lend Canadian mining operations legitimacy even where communities are terrorized by organized crime and there are public allegations that the company has a criminal organization on its payroll.

Furthermore, contrary to Champagne's version that Canada is helping other governments to strengthen weak legal and regulatory environments, the experiences in Colombia, Ecuador, and Honduras demonstrate Canada's willingness to intervene to lobby for detrimental laws and even to oppose the rule of law to facilitate the interests of globalized mining firms. While strengthening advantages for mining companies, the Colombian mining law jeopardizes the restitution of land for displaced peoples from mineral-rich lands in war-torn parts of the country, at the same time that it undermines the self-determination of Indigenous and Afro-descendant peoples and puts the livelihoods of small-scale miners at risk. The failure to apply a constitutional decree in Ecuador has put fragile water supplies and the rural and Indigenous people who depend on them at risk in the highlands and southern Amazon of Ecuador, allowing deep-seated conflicts to fester. The Canadian state lobby in Honduras undermined tenacious efforts by affected communities and civil society organizations to pressure authorities to put strict limits on any future mining in Honduras after having experienced the devastating impacts to water and public health in an area where a Canadian gold mining mammoth operated for less than ten years. If the government has been strengthening laws and regulations, it has been to put transnational corporations at greater advantage over the lives of current and future generations in Latin America.

Additionally, very much to the contrary of being able to rely on companies to play a constructive role when conflicts arise, there is a growing number of mining companies suing governments in Latin America where community opposition, environmental measures or efforts to reap greater benefits from mining have hit companies' bottom lines (Pérez Rocha and Moore 2019). Enabled by an expansive web of free trade and other international investor protection agreements, the Canadian government is a fierce defender of this legal framework through which Canadian mining companies are bringing suits for hundreds of millions and even billions of dollars to discipline governments for any measure that undermines their opportunities to profit exorbitantly from mineral extraction. In contrast, the Canadian government offers little more than voluntary, non-binding complaints mechanisms to workers and communities seeking to have their complaints heard. It is even plausible that the Canadian government may continue troubleshooting for those very same firms even while a complaint is in process.

Far from needing more Canada in the world and well beyond an issue for the "Canadian brand" or reputation, a serious reckoning is required to take stock of the intensifying violence and grave injustices being perpetuated through the globalized mining model in Latin America, which has been taking place with Canadian state support. Putting a halt to such support and taking significant steps to address the mounting social, economic, and ecological disasters that have been mounting alongside Canadian mining companies' super-profits is the only way that the Canadian flag will start to shake the neocolonial connotations it has earned as a tireless mining promoter in the region.

NOTES

1 Importantly, they note that this is an underestimation in actual flows since Statistics Canada does not calculate Canadian capital that is routed through Offshore Financial Centres.
2 O'Connor and Bohórquez (2010) cite 68 per cent of displaced peoples having come from mining zones, while Francisco Ramírez Cuellar estimates that 87 per cent of displaced peoples come from mining and energy areas (Vicente et al. 2011).
3 Access to information request A-2010–00758/RF1, pages 000213–14 and 000216.
4 In September 2010, Silvermex Resources Ltd. and Genco Resources agreed to combine their companies, the result of which was called Silvermex Resources Inc. In July 2012 First Majestic Silver purchased Silvermex and the La Guitarra mine along with it (Marketwired 2010; First Majestic Silver 2012).
5 Access to information request A-2010–00758/RF1, pages 000213–14 and 000216.

REFERENCES

Americas Policy Group. 2012. *Briefing Note: Mining.* (12 April). Ottawa: Canadian Centre for International Cooperation.
Anderson, Sarah, and Manuel Perez-Rocha. 2013. *Mining for Profits in International Tribunals: Lessons for the Trans-Pacific Partnership.* Washington, DC: Institute for Policy Studies.
Barrick. 2018. "Barrick Announces Departure of Kelvin Dushnisky." (23 July). Accessed 1 November 2021. https://barrick.q4cdn.com/788666289/files /press-release/2018/Barrick-Announces-Departure-of-Kelvin -Dushnisky.pdf.
CNCA (Canadian Network on Corporate Accountability). 2017. *Dirty Business, Dirty Practices: How the Federal Government Supports Canadian Mining,*

Oil and Gas Companies Abroad. Ottawa: Canadian Network on Corporate Accountability. Accessed 1 November 2021. http://www.halifaxinitiative .org/sites/halifaxinitiative.org/files/DirtyPractices.pdf.

– 2020. *Canadian Ombudsperson for Responsible Enterprise (CORE): Approach with Caution*. Ottawa: Canadian Network on Corporate Accountability. Accessed 1 November 2021. http://cnca-rcrce.ca/wp-content/uploads /2020/04/core-caution-E-1.pdf.

Contracorriente. 2017. "Después de 3 años declaran a lugar recurso de inconstitucionalidad contra Ley de Minería." (14 August). Accessed 1 November 2021. https://contracorriente.red/2017/08/14/despues-de -3-anos-declaran-a-lugar-recurso-de-inconstitucionalidad-contra-ley-de -mineria/.

Cumming, John. 2017. "In-depth Interview: Barrick's Dushnisky and Goldcorp's Garofalo on Gold Mining's Future." Interview by Greg Huffman. *The Northern Miner* (11 June). Accessed 1 November 2021. http:// www.northernminer.com/people-in-mining/depth-interview-barricks -dushnisky-goldcorps-garofalo-gold-minings-future/1003787114/.

Dawson, Laura. 2014. *Canada's Trade with Mexico: Where We've Been, Where We're Going and Why It Matters*. Ottawa: Canadian Council of Chief Executives. Accessed 1 November 2021. https://www.deslibris.ca /IDFR/242536.

DFTAD (Department of Foreign Affairs, Trade and Development Canada). 2013. *Global Markets Action Plan: The Blueprint for Creating Jobs and Opportunities for Canadians through Trade*. Ottawa – Ontario: DFTAD.

First Majestic Silver Corp. n.d. "Overview: La Guitarra Silver Mine." Accessed 1 November 2021. https://www.firstmajestic.com/projects/exploration -development/la-guitarra/.

– 2012. "First Majestic Completes Acquisition of Silvermex Resources." (2 July 2). Accessed 1 November 2021. https://www.firstmajestic.com /investors/first-majestic-completes-acquisition-of-silvermex-resources.

Flores, Ezequiel. 2016. "Nuevo Balsas, Guerrero, sitiado por el narco." *Proceso* (7 June). Accessed 1 November 2021. https://www.proceso.com.mx /reportajes/2016/6/7/nuevo-balsas-guerrero-sitiado-por-el-narco -165425.html.

Garibay Orozco, Claudio, and Alejandra Balzaretti Camacho. 2009. "Goldcorp y la reciprocidad negative en el paisaje minero de Mezcala, Guerrero." *Desacatos* 30 (2009): 91–110.

Global Affairs Canada. 2018. "The Government of Canada Brings Leadership to Responsible Business Conduct Abroad." (17 January). Accessed 1 November 2021. https://www.canada.ca/en/global-affairs/news/2018 /01/the_government_ofcanadabringsleadershiptoresponsible businesscond.html.

Goldcorp. 2016. *Annual Information Form for the Financial Year Ended December 31, 2015*. (29 March 29). Vancouver: Goldcorp.

Gordon, Todd, and Jeffery R. Webber. 2016. *Blood of Extraction: Canadian Imperialism in Latin America*. Halifax: Fernwood Publishing.

Imai, Shin, Leah Gardner, and Sarah Weinberger. 2017. "The 'Canada Brand': Violence and Canadian Mining Companies in Latin America." *Osgoode Legal Studies Research Paper* 17. Accessed 1 November 2021. http://dx.doi.org/10.2139/ssrn.2886584.

Inka Kola News. 2015. "The Torex (TXG.to) Can of Worms." (8 February). Accessed 1 November 2021. https://iknnews.com/the-torex-txg-to-can-of-worms/.

Kamphuis, Charis. 2018. "Canadian Economic Diplomacy: Policy Gaps, Human Rights Impacts & Recommendations." Submission to the United Nations Working Group on Business & Human Rights on behalf of the Justice & Corporate Accountability Project. Accessed 1 November 2021. https://ssrn.com/abstract=3125011.

Kirsch, Stuart. 2014. *Mining Capitalism: The Relationship Between Corporations and Their Critics*. Berkeley: University of California Press.

Mandujano, Isaín. 2012. "Represión al estilo Sabines." *Proceso* (24 November). Accessed 1 November 2021. https://vlex.com.mx/vid/represion-estilo-sabines-409705905.

Marketwired. 2010. "Silvermex Resources Ltd. and Genco Resources Ltd. Enter into Business Combination Agreement." (20 September). Accessed 1 November 2021. https://www.siliconinvestor.com/readmsg.aspx?msgid=26835504.

McSheffrey, Elizabeth. 2018. "At Long Last, a Human Rights Ombudsperson." *Canada's National Observer* (17 January).Accessed 1 November 2021. https://www.facebook.com/nationalobserver/videos/2019139771435456/.

MiningWatch Canada. 2012. "Canada's Subsidies to the Mining Industry Don't Stop at Aid." (15 June). Accessed 1 November 2021. https://miningwatch.ca/blog/2012/6/15/canada-s-subsidies-mining-industry-don-t-stop-aid-political-support-betrays.

– 2013a. "Honduran Mining Law Passed and Ratified, but the Fight is Not Over." (24 January).. Accessed 1 November 2021. https://miningwatch.ca/news/2013/1/24/honduran-mining-law-passed-and-ratified-fight-not-over.

– 2013b. "Backgrounder: A Dozen Examples of Canadian Mining Diplomacy." (8 October). Accessed 1 November 2021. https://miningwatch.ca/blog/2013/10/8/backgrounder-dozen-examples-canadian-mining-diplomacy.

– 2015a. "Honduran Organizations Fight to Have Canadian-Backed Mining Law Declared Unconstitutional." (26 February). Accessed 1 November 2021. https://miningwatch.ca/blog/2015/2/26/honduran-organizations-fight-have-canadian-backed-mining-law-declared.

- 2015b. "Ottawa We Have a Problem." (13 October). Accessed 1 November 2021. https://miningwatch.ca/blog/2015/10/13/ottawa-we-have-problem.
- 2016. "Two Years since Ayotzinapa: Mexico Is a Graveyard and Canada Is Quarrying for Headstones." (26 September). Accessed 1 November 2021. https://miningwatch.ca/blog/2016/9/26/two-years-ayotzinapa-mexico -graveyard-and-canada-quarrying-headstones.
- 2018. "Canadian Embassy in Mexico Subject of Complaint to Public Sector Integrity Commissioner." (5 February). Accessed 1 November 2021. https:// miningwatch.ca/blog/2018/2/5/canadian-embassy-mexico-subject- complaint-public-sector-integrity-commissioner.
- 2019. "Abarca Family Files with Federal Court of Appeal, Insisting that Canadian Embassy in Mexico Must Be Investigated." (19 August). Accessed 1 November 2021. https://miningwatch.ca/news/2019/8/19/abarca-family -files-federal-court-appeal-insisting-canadian-embassy-mexico-must-be.
MiningWatch Canada, CENSAT-Agua Viva, and Inter Pares. 2009. *Land and Conflict: Resource Extraction, Human Rights, and Corporate Social Responsibility: Canadian Companies in Colombia*, September 2009. Ottawa: Interpares. Accessed 1 November 2021. https://miningwatch.ca/sites /default/files/Land-and-Conflict.pdf.
MiningWatch Canada and ICLMG (International Civil Liberties Monitoring Group). 2015. "In the National Interest: Criminalization of Land and Environment Defenders in the Americas." Full Discussion Paper, August 2015.
MiningWatch Canada and United Steelworkers. 2015a. "Unearthing Canadian Complicity: Excellong Resources, the Canadian Embassy, and the Violation of Land and Labour Rights in Durango, Mexico." Accessed 1 November 2021. https://miningwatch.ca/sites/default/files/excellon _report_2015-02-23.pdf.
- 2015b. "Government Documents Reveal Canadian Embassy Backed Mining Abuses in Mexico." (25 February). Accessed 1 November 2021. https://miningwatch.ca/news/2015/2/25/government-documents-reveal -canadian-embassy-backed-mining-abuses-mexico.
Moody, Roger. 2007. *Rocks and Hard Places: The Globalization of Mining.* London: Zed Books.
Moore, Jen, Robin Broad, John Cavanagh, René Guerra Salazar, Meera Karunananthan, Jan Morrill, Manuel Pérez-Rocha, and Sofía Vergara. 2014. "Debunking Eight Falsehoods by Pacific Rim Mining/OceanaGold in El Salvador." (17 March). Accessed 1 November 2021. https://ips-dc.org /debunking_eight_falsehoods_by_pacific_rim_mining/.
Moore, Jennifer, and Teresa Velásquez. 2012. "Sovereignty Negotiated: Anti- mining Movements, the State and Multinational Mining Companies under Correa's 21st Century Socialism." In *Social Conflict, Economic Development*

and Extractive Industries: Evidence from South America, edited by Anthony Bebbington, 112–33. London: Routledge.

Natural Resources Canada. 2021. "Canadian Mining Assets (CMA) by Country and Region, 2018 and 2019." Last Modified 2 February. Accessed 1 November 2021. https://www.nrcan.gc.ca/mining-materials/publications /15406.

O'Connor, Dermot, and Juan Pablo Bohórquez Montoya. 2010. "Neoliberal Transformation in Colombia's Goldfields: Development Strategy of Capitalist Imperialism?" *Capital and Society* 43, no. 2: 85–118.

Otros Mundos AC/Chiapas – Amigos de la Tierra México. 2010. *Mexico a Cielo Abierto*. Directed by Andres DCO. Accessed 1 November 2021. http:// archive.org/details/MexicoACieloAbierto_124.

Pérez-Rocha, Manuel, and Jen Moore. 2019. *Casino Extraction: Mining Companies Gambling with Latin American Lives and Sovereignty through International Arbitration*. Washington, DC: Institute for Policy Studies.

Procuraduria General de Justicia del Estado Mexico. 2009. "Public Attorney's Initial Assessment of Blackfire Complaint against Mariano Abarca (Averiguación Previa Número 00033/FS10/2009)." Chiapas: Procuraduria General de Justicia del Estado, Fiscalía de Distrito Fronterizo Sierra, Fiscalia del Ministerio Público de Chicomuselo.

Rivera Sierra, Enrique. 2009. "Mariano Abarca en la embajada canadiense, 22 JULIO 2009." YouTube video, 1:56, posted on 21 August. Accessed 1 November 2021. http://www.youtube.com/watch?v=zwGavLzTob8.

Secretaría de Economía México. Dirección General de Desarrollo Minero. 2018. "Diagnóstico de Empresas Mineras Mexicanas con Capital Extranjero en la Industria Minero Metalúrgica del país." First quarter of 2018. Accessed 1 November 2021. https://www.gob.mx/cms/uploads /attachment/file/398126/Diagnostico_1er_Semestre_Estadisticas_2018.pdf.

Szablowski, David. 2007. *Transnational Law and Local Struggles: Mining, Communities and the World Bank*. Oxford: Hart Publishing.

The Globe and Mail. 2018. "Globe Editorial: Ottawa Needs to Give New Human Rights Watchdog Actual Teeth." (21 January). Accessed 1 November 2021. https://www.theglobeandmail.com/opinion/editorials/globe-editorial -ottawa-needs-to-give-new-human-rights-watchdog-actual-teeth /article37672114/.

United Steelworkers, Common Frontiers, and MiningWatch Canada. 2010. "Report from the March 20–27, 2010 Fact-Finding Delegation to Chiapas, Mexico to Investigate the Assassination of Mariano Abarca Roblero and the Activities of Blackfire Exploration Ltd." Accessed 1 November 2021. http:// s3.amazonaws.com/isuma.attachments/5073_Chiapas_delegation_report _2010.pdf.

– 2013. "Corruption, Murder and Canadian Mining in Mexico: The Case of Blackfire Exploration and the Canadian Embassy." (May). Accessed 1 November 2021. https://miningwatch.ca/sites/default/files/blackfire_embassy_report-web.pdf.

Vicente, Ana, Neil Martin, Daniel James Slee, Moira Birss, Sylvain Lefebvre, and Bianca Bauer. 2011. "Minería en Colombia: a qué precio?" *PBI Colombia,* Boletín Informativo No. 18, November.

Villagomez, Enrique. 2016. "Firma canadiense invierte 800 mdd en nuevo desarrollo minero en Guerrero." *El Financiero* (28 April). Accessed 1 November 2021.http://www.elfinanciero.com.mx/empresas/firma-canadiense-invierte-800-mdd-en-nuevo-desarrollo-minero-en-guerrero.html.

Voices-Voix. 2017. "Excellon Resources, the Canadian Embassy and Ejido Community in Mexico." (5 July).

Weisbart, Caren. 2018. "Diplomacy at a Canadian Mine Site in Guatemala." *Critical Criminology.* DOI: 10.1007/s10612–018–9422-y. Accessed 1 November 2021. https://www.researchgate.net/publication/328174105_Diplomacy_at_a_Canadian_Mine_Site_in_Guatemala.

World Gold Council. n.d. "Conflict-Free Gold Standard." Accessed 1 November 2021. https://www.gold.org/about-gold/gold-supply/responsible-gold/conflict-free-gold-standard.

10 Voluntary or Legislated? The Home-Country Regulation of Canadian Mining Companies in Latin America

PAUL ALEXANDER HASLAM

The promotion and protection of Canadian economic interests abroad through a web of trade and investment agreements was the corner-stone of Stephen Harper's Americas Strategy and has remained so for the Trudeau government. In concrete terms, this has been a policy about Canadian mining companies, which are the most important Canadian investors in the region. However, Canadian mining companies operating in Latin America have also attracted considerable attention from academics and activist organizations alleging human rights abuses and spurring a public debate about how the Canadian government should regulate the behaviour of these firms abroad. In this regard, the government's policy response to this problem, which focused on encouraging the adoption of high standards of corporate social responsibility (CSR) by mining firms, should be considered an integral part of the Americas Strategy from Harper through Trudeau, even if not formally named as such.

The purpose of this chapter is to examine the government's CSR policy, through an overview and analysis of the claims and proposed solutions made by both sides of the debate about how to regulate Canadian mining companies abroad. In broad terms, the debate has settled into two dichotomous recipes. Both approaches show some agreement on what should be the standard for acceptable behaviour – a set of internationally recognized standards of CSR.[1] However, activists have argued for a legislated solution, which, in its most common formulation, would *require* companies to adhere to and respect a list of CSR and human rights codes, under threat of sanction. In contrast, mining companies and their representatives, for the most part, have argued for the *voluntary* application of responsibility standards. Activists see voluntary solutions as ineffective, while industry sees legislated solutions as costly and unresponsive to change. In my discussion of these options, I ask how each is likely to affect the drivers of social conflict around mining developments and corporate accountability, and on that basis, I offer an assessment of the government's CSR policy as an unofficial pillar of Canada's Americas Strategy.

Complex Causes, Unclear Solutions

The mining sector is a unique business sector in Canada. It is the source of much of Canada's foreign direct investment (FDI) abroad, in a country that has a very poor record of corporate internationalization beyond the United States. Mining is undeniably a leading sector in which Canadian firms are globally competitive and contribute significantly to employment (375,000 people), technological progress, and national income (7 per cent of GDP) at home (UN Working Group 2018, 14). The role of Canadian stock market exchanges, particularly the Toronto Stock Exchange –TSX – (but also the Vancouver Stock Exchange), in raising capital for small junior companies is further evidence of the country's global competitive advantage in mining (Deneault and Sacher 2012, 16–17). Canadian mining investment is particularly important in Latin America, where, it is estimated by the Mining Association of Canada, that Canadian mining companies account for $78 billion in foreign investment abroad, and 40 per cent of all outward FDI (Gratton 2017).

However, the wave of mining investment abroad since the early 1990s has also been associated with social conflict with communities proximate to mines and mining projects in Latin America and elsewhere in the developing world – although it should be recognized that not all mines or mining projects are linked to conflict. Haslam and Ary Tanimoune (2016), using activist sources, find that the proportion of mines characterized by "known conflicts" across their sample of 783 properties in 23 countries of Latin America is 24.4 per cent. The causes of mine–community social conflicts are complex but can be summarized as big impacts, big rents, and weak local institutions (for an overview, see Conde and Le Billon 2017). On the first point, case studies have pointed to tensions created by the opening of a new extractive frontier near communities with little experience with mining and the use of open-pit techniques and toxic chemicals to exploit low-grade disseminated deposits, which can make large and visible impacts on the environment and which are often viewed by peasants as being incompatible with existing agricultural livelihoods (Bebbington et al. 2008; Bebbington 2012; Walter and Martinez-Alier 2010).

A second strand of literature points to distributive issues related to how rents (or benefits) are distributed to affected communities, as social groups struggle over who gets what and how to get more from the company (Arellano-Yanguas 2011). Distributional conflicts point to problems that can be caused when modern high-tech mines only provide limited benefits in terms of community projects, the subcontracting of local enterprise and direct employment benefits to locals (Humphreys Bebbington and Bebbington 2010, 262–5). The distributional perspective is also supported by evidence that local poverty is associated with social conflict (Haslam and Ary Tanimoune 2016, 413; Ponce and McClintock 2014, 131). Thirdly, recent work has pointed to the role of weak host country institutions – especially at the local level – which often fail to effectively represent or channel citizen concerns, causing protest to spill into the street (Amengual 2018; Jaskoski 2014, 897; Speigel 2012, 201). Weak local institutions are also at the centre of activist concerns that mining companies are unlikely to be held accountable in host-country legal systems. Evidence for these three hypotheses in Latin America is extensive, based on individual case studies, and more recently, large-n statistical analyses, some of which are cited above.

Activist Claims

A fourth causal hypothesis has emerged in Canada, associated with the "Imperial Canada" literature and publicized by activist organizations, which suggests that the home-country policy regime shapes corporate behaviour. This literature has tended to make two related claims: firstly, that Canadian companies have been particularly associated with problems abroad, in comparison to other nationalities; and secondly, that Canada's home-country regulatory regime is "ultra-permissive," with "virtually no supervision of their potential wrongdoing beyond Canada's borders," or legal accountability at home (Deneault and Sacher 2012, 12, 51). Activists argue that by failing to adequately regulate mining companies at home, the Canadian government has been "complicit" in human rights abuses by Canadian companies in foreign countries.

Specifically, Canadian governments (including the provinces) are thought to facilitate bad corporate behaviour abroad through weak stock market disclosure rules (Deneault and Sacher 2012, 16–17; Imai et al. 2017, 40), limited legal options in Canada for foreign victims of companies (CNCA 2007, 5.1), state-sponsored investment financing and insurance, and the diplomatic promotion and defence of Canadian mining firms (Campbell 2010, 199; CNCA 2007, 4.1, 6.1; Veltmeyer 2013, 85–6; see also chapter in this volume by Moore). If, as many activists and academics believe, Canadian mining firms behave particularly badly, and a weak legal/regulatory regime at home (combined with active diplomatic promotion abroad) is the cause of problems, then, it is argued, the government has a legal and moral responsibility to increase home-country regulatory oversight of company operations abroad (CNCA 2007; DPLF 2014; Imai 2017).

Evidence for these arguments has been provided by a diverse set of NGOs working on the sector, as well as through a series of reports that aimed to enumerate the problems. A CCSRC (Canadian Centre for the Study of Resource Conflict) report commissioned by PDAC in 2009, and subsequently leaked to the public, identified 171 incidents of social conflict worldwide and reported that Canadian firms were implicated in a plurality of these incidents, "tripling its closest peer, Australia" (CCSRC 2009, 16). This report was widely and approvingly cited by activists, although its evidence showed that on a proportional basis (34 per cent of conflicts; but 75 per cent of the companies) Canadian firms were less likely to be associated with social conflict than firms of other nationalities (CCSRC 2009, 7). Subsequent reports by the Working Group on Mining and Human Rights in Latin America, an alliance of seven regional NGOs, and Shin Imai's "Canada Brand" report also sought to count and categorize various human incidents "associated" with Canadian firms in Latin America (DPLF 2014, 8; Imai 2017).

Such incidents reported by activist organizations are of great concern and demand a response – but what response? The causal claim that began this section, that linked a weak home-country regulatory environment to human rights incidents or bad behaviour abroad has not received rigorous evidentiary support, unlike the other locally grounded hypotheses discussed above. A recent analysis, which contextualizes the data provided by activist organizations about social conflicts within the larger universe of mining projects of any nationality in the five largest mining countries of Latin America (Argentina, Brazil, Chile, Mexico, and Peru) showed that Canadian mining firms had a 21 per cent likelihood of being involved in a social conflict with nearby communities, as compared to a 28 per cent likelihood for foreign but non-Canadian firms, and a 7 per cent likelihood for local firms (Haslam, Ary Tanimoune, and Razeq 2018, 545).[2] If Canadian mining firms perform *better* than non-Canadian foreign firms, then it is unlikely that the home-country regulatory regime, generally portrayed as weaker than other mining home countries in the literature, is the cause (Haslam, Ary Tanimoune, and Razeq 2018, 546).

Notwithstanding the limited evidence for a causal link between weak regulation at home and poor performance abroad, there remains a moral argument that the home country and its citizens who benefit from the mining industry have an obligation to ensure the ethical conduct of companies abroad. At a minimum, the home country should not overlook its human rights obligations if promoting the interests of mining firms (see Moore, this volume). Since 2011, this intuitive stance has dovetailed with the objectives of the UN *Guiding Principles on Business and Human Rights*, which underline the obligation of states to ensure that third parties, including companies, respect human rights.

Industry Claims

The academic and activist literature has shown much less interest in the CSR actions of Canadian mining firms. As previously mentioned, these observers have tended to view CSR practices cynically, seeing them as epiphenomenal public relations ploys, sign of corruption, and contributing to the division of communities (Clark and North, 2006). In this regard, only a few authors have taken CSR seriously as a potential contributor to reducing conflict and improving firm–community relations (Sagebien et al. 2011; Dashwood 2012; Webb 2012; Conde and Le Billon 2017; Haslam 2018; Haslam, Ary Tanimoue and Razeq 2018; Haslam 2021).

In fact, there has been a progressive ratcheting-up of voluntary standards of behaviour and accountability over the last fifteen to twenty years, in significant part led by leading companies, industry associations, and international organizations. For example, Dashwood, who examines the emergence of global norms of "sustainable mining" in the industry, points to the early role of leading Canadian firms. Dashwood presents compelling evidence that some leading mining firms were proactive about shaping CSR norms, promoting higher standards of corporate behaviour, engaging with NGOs, and building international institutions to promote these norms globally (Dashwood 2007, 131–2, 147–50).

Whereas CSR in mining began with vague, general statements about corporate ethics and stakeholder engagement (a state of affairs that still exists in many small companies), it has increasingly evolved into specific and measurable commitments requiring extensive collection of data, public reporting, and third-party verification or assurance. Industry has tended to see the adoption, promotion, and further development of CSR codes by government and business associations as appropriate to the task of addressing allegations associated with social conflict. A consultant's report prepared for the Mining Association of Canada's CSR committee suggested that, increasingly, requirements associated with project financing, act to "harden" CSR codes (Fasken Marineau 2012, 75).

In addition to the many critical studies cited above, cases from the region also provide numerous examples of CSR "working" to limit or reduce social conflict and achieve developmental goals. For example, Jaskoski (2014) highlights the successful dialogue process at the Quellaveco mine in Peru; Martinez and Franks (2014) show investments in human capital to improve perceptions of a firm in Chile; Amengual (2018) provides strong evidence that CSR provides public goods to communities in Bolivia; and Haslam (2021) examines how CSR reduces conflict in eight cases in Argentina and Chile. In some instances, improved CSR contributions may be brokered by activist states, as in Peru's famous "voluntary contributions" programme, and the trust funds created in the Argentine provinces of San Juan and Santa Cruz (Haslam 2018).

In September–October 2017 the Parliamentary Subcommittee on International Human Rights held hearings on issues related to human rights issues surrounding natural resource extraction in Latin America. President and CEO of the Mining Association (MAC) of Canada, Pierre Gratton, outlined the views of MAC regarding the problem of social conflict and alleged human rights abuses in the region, which he recognized as troubling, but as outside the norm of responsible business practice by Canadian mining companies. Gratton noted the success that Canada and the Mining Association of Canada had had with voluntary measures – particularly MAC's own standard, *Towards Sustainable Mining* (TSM), which is a CSR and reporting standard that is mandatory for members of the business association. Additionally, Gratton pointed to the potential consequences for mining companies, since 2014, resulting from the Canadian government's "enhanced" CSR policy, namely, the withdrawal of diplomatic support from the trade commissioners' service and access to Export Development Canada (EDC) financing and risk insurance (Gratton 2017).

In this regard, the industry position points to the utility and effectiveness of voluntary codes of conduct, accepts the organic development of domestic court rulings on jurisdiction over incidents alleged to have been caused by Canadian firms abroad, and supports the use of a cooperative model of dispute mediation and joint fact-finding for the purpose of reducing tensions and social conflict at mining sites (Gratton 2017).

Government Response

Since the debate on what to do about the behaviour of Canadian mining companies abroad has emerged, the policy of the Canadian government has evolved from initially siding with industry on the promotion and voluntary adoption of standards for responsible behaviour abroad to include some punitive disciplines for non-conformity with these standards, culminating in a new policy initiative (January 2018) to create an Ombudsperson's office with investigative powers. This evolution is seen in three CSR policies that I have characterized as an unofficial pillar of the Americas Strategy: *Building the Canadian Advantage* (2009); *Doing Business the Canadian Way* (2014); and the creation of the *Canadian Ombudsperson for Responsible Enterprise - CORE* (2018–19).

The National Roundtables on Corporate Social Responsibility and the Canadian Extractive Industry in Developing Countries were convened by the Liberal government in 2006. The multistakeholder Advisory Group that led the process (involving civil society, industry, and academic representatives) issued a consensus report in 2007 that recommended the following: the establishment of Canadian CSR standards based on existing international standards (then the IFC Performance Standards and "Voluntary Principles"), reporting based on the Global Reporting Initiative (GRI), the creation of an Ombudsperson with investigative powers, and the establishment of a Compliance Review Committee to evaluate company compliance and recommend the withdrawal of governmental support for companies in the case of non-compliance (Advisory Group 2007, iii; Keenan 2013, 112–14). At the time, the recommendation to create an independent Ombudsperson with investigative powers was considered a key victory by civil society organizations.

In 2009, under the Harper government, these recommendations were translated into policy, titled *Building the Canadian Advantage (BCA)*. From the perspective of civil society, the policy was a disappointment, as it failed to materialize two key Roundtables recommendations: mandatory standards with consequences for non-compliance and an Ombudsperson with investigative powers (Keenan 2013, 114). Instead, *BCA* promoted the voluntary adoption of three CSR and reporting standards over and above the existing OECD Guidelines (IFC Performance Standards, Voluntary Principles, and GRI), established a small fund to promote CSR and extractive sector governance abroad, established a CSR Centre of Excellence to promote best practices among firms, and created the Office of the Extractive Sector CSR Counsellor (DFAIT 2009). The CSR Counsellor was the most institutionally innovative and controversial part of the strategy, intended to substitute for the Roundtables' recommendation for an extractive sector Ombudsperson. The mandate of the Counsellor, however, was principally to promote CSR to mining companies, and it had no independent investigative powers. The office could receive requests to review social conflicts from any "individual, group or community that reasonably believes that it is being or may be adversely affected by the activities of a Canadian extractive sector company," and encourage mediation, but only with the consent of both parties (DFAIT 2009). In practice, participation in the review and mediation process was frequently and equally rejected by corporations and complainants.

Close on the heels of this disappointment for civil society organizations, Liberal MP John Mackay was able to take advantage of the minority parliament to advance a private member's bill that offered to convert the voluntary undertakings of *Building the Canadian Advantage* into mandatory legislated commitments with Bill C-300 (of the 40th parliament), the *Corporate Accountability of Mining, Oil and Gas Corporations in Developing Countries Act*. Key provisions of Bill C-300 included a legal requirement to comply with international standards (OECD Guidelines, IFC Performance Standards, the Voluntary Principles, and international human rights law) to be eligible for government support; government support could be withdrawn for non-compliance; and the establishment of a complaints mechanism in which good faith complaints were to be investigated by the Minster of Foreign Affairs and International Trade, whose findings on compliance could be published. Bill C-300 was seen as a "strategic starting point" by civil society (Keenan 2013, 114). However, the bill narrowly failed to pass the third reading in October 2010, due to opposition from the Conservative government, and important defections on the Liberal, Québécois Bloc and New Democratic benches (Keenan 2013, 116). Two similar efforts by the NDP failed on first reading (MAC 2012, 18).

In November 2014 *Building the Canadian Advantage* was updated by the Harper government as *Doing Business the Canadian Way*, more commonly known as the *Enhanced CSR Strategy*. The Enhanced Strategy added the UN *Guiding Principles on Business and Human Rights*, and OECD Due Diligence Guidance for Responsible Supply-Chains of Minerals from Conflict-Affected and High-Risk Areas to its benchmark standards. The language of the new policy was assertive, suggesting that in the case of doubt, the government "expected" companies to comply with the "higher, more rigorous standard" (DFAIT 2014, 3, 6).

Most importantly, it established a disciplinary function in the CSR Counsellor's Office, in which companies who refused to participate in Counsellor-recommended dispute-resolution processes, could face the withdrawal of "enhanced" diplomatic support. Company representatives were to sign an "integrity declaration" to receive "enhanced" support from foreign trade mission, and it was on the basis of violation of this contract that governmental support could be withdrawn (UN Working Group 2018, 9). Non-compliance with the policy would be considered by Export Development Canada in its due diligence on corporate applications for financing and risk insurance (DFAIT 2014, 11–12). Whether these incentives were sufficient to force unwilling companies to the dialogue table remained unclear. In the six cases considered by the NCP since the adoption of the new rules (2014–18), only one company refused to participate, a better success rate than previous years. That company, China Gold International Resources Corp., was disciplined with the withdrawal of enhanced support.

In the run-up to the federal election of October 2015, the campaigning Liberals promised the Canadian Network on Corporate Responsibility, an NGO coalition, that it would set up an "independent Ombudsperson office" with investigative powers (Munson 2017). On 17 January 2018 a new policy for a Canadian Ombudsperson for Responsible Enterprise (CORE) was announced. The CORE is inspired by the *UN Guiding Principles on Business and Human Rights*, which underlines the state's responsibility to promote respect for international human rights obligation to third parties (such as companies), and the firms' obligation to respect human rights and to conduct due diligence on potential human rights risks and remedy problems (Ruggie and Sherman 2017, 923–4). More generally, the creation of the Ombudsperson fulfils a recommendation of the Roundtables for a body with investigative powers, the ability to make recommendations, including that non-complying firms face governmental sanction – principally the loss of "enhanced diplomatic support" – and publicly report on its finding. While an Ombudsperson was appointed in April 2019, civil society groups remained unsatisfied with its development through summer 2020, suggesting that its protocols regarding admissibility, engagement, and investigation were inappropriate to substantively addressing human rights violations (CNCA 2020; also see the chapter by Deonandan and Schmuland in this volume).

How Effective Is Self-Regulation?

Voluntary and legislative alternatives for managing the extractive industry abroad are often presented, respectively, as soft and ineffective constraints, as compared to hard and effective obligations. This misrepresents the nature of modern CSR in the extractive industry, which is often "hardened" through integration with incentives that fit with the business logic of the firm and overstates the willingness or ability of home-country governments to discipline their private sectors abroad.

The effectiveness of CSR and voluntary codes or standards has attracted significant attention in international business studies and international relations. The rise of CSR is partially related to the withdrawal of the state from direct interference or regulation of various economic sectors associated with the return to liberal economic policies in the 1990s. But it is also related to recognition by state actors (and international organizations) that, in the era of globalization, many subjects are not easily regulated by states and that self-regulation has a role to play in covering a "governance gap" and the provision of public goods when the state fails or is unable to do so (Frynas and Stephens 2015, 483). The management of ethical issues across borders is one of those governance gaps.

The business literature has struggled with proving the business case for CSR. However, following a brilliant study by Henisz et al. (2014) that linked good stakeholder relations with higher market valuation, the literature increasingly finds that mining companies have a material interest in CSR policies that reduce social risk and the costs associated with it, including citizen blockades, destruction of property and violence, or drawing the disciplinary attention of political authorities. But effective CSR also depends on the specific nature of the CSR codes adopted, and the broader political context and implementation. In contrast to the vague or aspirational codes of behaviour that have been correlated with weak implementation (Sethi and Emelianova 2006, 233), Kolk et al. show that the "compliance likelihood" of a company increases when its CSR commitments are highly specific, measurable, include reporting requirements, and are subject to verification (assurance) by an independent third party (Kolk, van Tulder, and Welters 1999, 167). These criteria have been "mainstreamed" in leading CSR approaches. Insofar as certain voluntary codes also permit sanctions for non-compliance (such as the withdrawal of financing or risk insurance, or expulsion from the membership group), it may be expected that these consequences "harden" voluntary undertakings in a way that effectively makes them mandatory (Fasken Martineau 2012, 8, 75).

By now, it is widely accepted that firms have diverse motivations for pursuing self-regulation (reducing risk, improving reputation, employee satisfaction, institutional learning, and competitiveness). Furthermore, as Haufler points out, even if states are not directly involved in regulating firms, the "shadow of the state," the fear of future regulation, is an important motivating factor (Haufler 2001, 20–6). Complementary, high social, and environmental regulatory standards in host states establish a floor for corporate behaviour but also raise the bar for companies seeking "social licence" to operate (Sagebien et al. 2011, 112). In this regard, effective regulation in the host country is viewed as the foundation for effective CSR.

However, lack of enforcement of legislation defines the weak states and the even weaker subnational jurisdictions of Latin America, as well as many other areas of the developing world. In these contexts, powerful actors are able to violate both the letter and the spirit of the law without consequences. The host state itself is often more interested in capturing the rents generated by mining (royalties, taxes), than supporting the demands of citizen-protestors (Haslam and Heidrich 2016). In this context, voluntary commitments of firms can, in fact, substitute for the rule of law – whether that outcome is viewed as philosophically desirable or not (Haslam 2018). Therein lies a paradox: In developing countries where low-quality institutions are least likely to provide a floor for building voluntary initiatives that go above and beyond the law, corporate social responsibility may be the best and only option for bridging governance gaps.

Despite the attention to home-country regulation by activists and academics, it is not clear that Canadian-based regulation is particularly important to the behaviour of companies abroad, especially considering the business-case drivers behind CSR and the myriad local factors, already discussed in this chapter, that have been proven to influence social conflict and which, due to proximity (and the logic of Occam's razor), are more likely to be causally significant. Nonetheless, it does seem likely that the threat of home-country regulation that has existed in Canada since 2007 has added urgency and seriousness to the pursuit of self-regulatory solutions by mining companies and their associations.

The Potential of CSR

How effective have CSR codes been in managing social conflict and human rights allegations in Latin America? We must admit, firstly, that the effect and effectiveness of the Canadian government's CSR strategy since 2009 cannot be rigorously evaluated with existing and publicly available data and, secondly, that there is very little generalizable academic research on this topic, beyond anecdotal "evidence." In this regard, we offer some hypotheses on the use of CSR codes and benchmarks but ultimately are limited in our ability to generate solid conclusions.

Additionally, the record on mediations initiated by the CSR Counsellor's Office and National Contact Point (NCP) is not encouraging. Since the creation of the CSR policy in 2009 several "reviews" have been undertaken related to social conflicts between individuals or communities in Latin America and Canadian mining companies, both by the CSR Counsellor (five of six reviews, 2009–13), and the National Contact Point (two of twelve specific instances, 2009–18). As previously noted, sanction (withholding of "enhanced" diplomatic support) has been an option available since 2014 (potentially covering six of these cases) – but had only been applied to one firm at the time of writing. Of the seven mining cases in Latin America reviewed by the Counsellor or NCP (2009–18), as a first step towards mediation, three were dropped because the company withdrew from participating, three were dropped because the Notifiers (complainants) withdrew from participating or could not be reliably contacted by the Office, and one case was dropped because the CSR Counsellor's Office determined it was without merit. In this regard, six of the seven cases "failed" due to an unwillingness of one of the parties involved to participate in mediation. This suggests that mediation is unlikely to resolve issues that are perceived by at least one party as a zero-sum game.

Nonetheless, it is clear that the use of CSR codes, and their intersection with microeconomic incentives for better behaviour, has dramatically increased over the last decade. Although a NRCAN evaluation of *Building the Canadian Advantage* found that neither the government's policy nor the specific codes of conduct it promoted received much attention in mandated stock market filings by companies, it did conclude that "more companies perceive voluntary CSR activities as being material" – that is, of significance to shareholder value, and requiring inclusion in company reports and filings (NRCAN 2013, 9, 59). The evidence compiled by NRCAN showed a steady increase in use of social and environmental keywords over time since 2004, with a steep jump in the use of "human rights" since 2009, although the most commonly cited terms in Latin America were related to the "environment," followed by "labour," and "governance and ethics" (NRCAN 2013, 46). While there appears to be increasing concern for CSR concepts by mining companies, the methodological weaknesses of the NRCAN study, namely, the lack of a non-Canadian comparator group, means these changes cannot be definitively attributed to home-country government policy.

It looks like distributive issues, namely, who gets what from a company's CSR programmes, or labour disputes, are more amenable to "management" than socio-environmentally based conflicts. For example, the previously cited study by Haslam, Ary Tanimoune, and Razeq, provided evidence for the greater success of Canadian mining firms, as compared to foreign non-Canadian firms, in poorer areas normally thought to be associated with distributional pressures – which they hypothesized to be due to better CSR programmes (2018, 542). In this regard, the promotion of dialogue and mediation between the company and complainants, which is at the heart of the voluntary regime, may offer the potential to address distributional drivers of social conflict.

For example, the 2011 Excellon case (Mexico) was an example of distributive demands (a labour dispute) in which better communication and structured dialogue facilitated by the CSR Counsellor's office could have been important – if the company had not dropped out of the process. However, the company saw the process as bogus and as having "legitimized illegitimate actors" (MAC 2012, 60).

Some stakeholders in the Roundtable process questioned the utility of the voluntary CSR Counsellor's Office, seeing it as "incongruous with principles of accountability and access to remedy," and unsure of its utility to affected communities (MAC 2012, 60). However, Notifiers (complainants) were as likely to drop out of sponsored mediation as companies, suggesting that the demands of community actors frequently could not be accommodated within such a framework for dialogue. More fundamentally, where the drivers of social conflict are concerns over risks to the environment and agricultural livelihoods, which local actors often see as zero-sum games, dialogue is unlikely to be viewed as productive by those same actors.

Would Activist Allegations be Actionable under a Legislated Solution?

If we admit that CSR codes, and voluntary approaches more generally, are limited in their ability to fully resolve the issues that underlie social conflicts around mines and mining projects, except when distributive in nature, we should similarly question the utility of legislating CSR commitments. Indeed, the most important, and the most difficult, question to answer is whether the allegations identified by academic and activist organizations would be actionable under a legislated solution in a way that is substantially more effective than the voluntary approach. Obviously, that depends on the details of the legislation (or the Order in Council governing the Ombudsperson's office), with the draft law proposed by the Canadian Network on Corporate Accountability (CNCA), representing an implausibly high watermark for legislation in which almost any actual or possible "harm" resulting from a project or proposed project could trigger investigation and sanction (CNCA 2016). This chapter will examine the more likely scenario reflected in the Roundtable recommendations, Bill C-300, and currently available information about the CORE.

Firstly, it needs to be recognized that the disciplinary actions initially envisaged by the Roundtables' Advisory Group, and now the CORE, are relatively mild. In the worst-case scenario of non-compliance with the standard (or non-participation in the process) and no (future) remedial action, the Ombudsperson could recommend the retraction of enhanced diplomatic support for the company, including access to EDC financing/risk insurance (Advisory Group 2007: v; GAC 2019). Realistically speaking, participation in a process (mediation) could be sufficient to satisfy the requirement of compliance. Similarly, it must be recognized that corporate action on remediation could be judged to be satisfactory (against the standards of the CSR codes) largely on procedural grounds, without addressing all substantive activist concerns. This should be particularly clear with regard to the implementation of certain standards promoted by activists such as "free, prior and informed consent" (FPIC) of Indigenous and other communities for mining projects, which is not (yet) widely accepted by national or international law (where substantive consultation is generally viewed as sufficient).

For example, in the Corrientes Resources case (Ecuador), the NCP found that the human rights claim of the Notifiers, based principally on lack of consultation with Indigenous and non-Indigenous groups, was without substance or merit. It was noted that the "OECD Guidelines do not include a requirement for free, prior and informed consent" and the Notifiers had already lost court cases on these issues in Ecuador. Concerns regarding complicity with state repression were similarly rejected by the NCP, as the Notifiers failed to clearly link these alleged actions with the company, and "applying the OECD guidelines to state actions is beyond the mandate of the Canadian NCP." Concerns over environmental impact and working conditions were also viewed to be unsubstantiated (Canadian NCP 2014, Sec. 16). These issues, particularly FPIC and alleged complicity with state repression, are the bread and butter of activist work on Canadian mining abroad and underlie many of the incidents identified in reports like Imai's "Canada Brand." The ruling of the NCP on Corrientes Resources suggests that key issues (from an activist perspective) may not be actionable under *either* a voluntary *or* a legislated regime benchmarked to CSR standards. But, of course, this depends on the standard. The gradual development of an FPIC standard for Indigenous communities in Canada through the courts and the Trudeau government's steps to operationalize the UN Declaration of Rights of Indigenous Peoples (UNDRIP) do raise the question of whether FPIC could emerge as a human rights standard that the CORE would be obliged to uphold – which really could be a game changer.

Under the *Guiding Principles*, a company's substantive obligation is to provide remedy "commensurate with, and proportional to, its involvement in the harm" (Ruggie and Sherman 2017, 927). In the case of directly causing or contributing to the non-respect of human rights, the company should respond through some kind of complaints mechanism (judicial or *corporate*). However, where causation is less directly tied to the company but where a business relationship (for instance, to a subcontractor) links the company to a problem, the "responsibility to respect human rights does not require that the enterprise itself provide for remediation" (Ruggie and Sherman 2017, 927). In this regard, the ability of the CORE to sanction companies with reference to the *Guiding Principles*, will depend on complex and subtle interpretations of the responsibility a company has for a particular event (caused, contributed, or linked by a business relationship to a responsible entity) – a level of nuance that has generally been absent from the widely publicized reports, cited above, alleging human rights abuses. Additionally, the ability of the CORE to act at all will depend on the will of the office holder, how it is funded, the powers delegated to it, whether it is independent from the government, and how it is overseen by its potentially fractious multisectoral Advisory Body.

Perhaps the most important aspect of the CORE and its elevation of the *Guiding Principles* as Canadian policy, is that it effectively creates an obligation for companies, who wish to avoid sanction, to conduct due diligence on human rights risks abroad. This obligation has the potential, via the analysis and reporting process, to start conversations in corporate and subsidiary offices that could lead to important preventive steps to minimize the occurrence of human rights impacts. Whether this mechanism is voluntary or legislated at the home-country level may not be consequential – and indeed the logic behind its intended effect is like that of other voluntary instruments. As a policy tool it clearly builds on the environmental impact assessments (often legislated by host countries but delegated to corporations), and social impact assessments (usually voluntary) that are widely practised by corporations. It is not the disciplinary threat to the company that is most important here; instead it is the fact that companies will create processes to conduct these assessments and consider internally how they can mitigate and remediate human rights impacts, which may have the greatest impact for people living proximate to mining projects. Yet, we should recall that this is not the same as people being able to claim their rights.

Another important consideration is the extent to which those affected by the actions of Canadian corporations are likely to be able to find redress or hold corporations accountable in the Canadian court system. Activist and academic observers have generally been critical of the limited legal opportunities afforded to foreign plaintiffs in Canada and have advocated for better legislated access, as has a UN Human Rights Council Working Group report on Canada (Deneault and Sacher 2012, 54–5; UN Working Group 2018, 18). Canadian courts have been reluctant to hear cases that originate in alleged wrongdoing abroad (excepting bribery and corruption, which are covered by Canadian criminal law) for a number of reasons: the principle of *forum non conveniens*, that cases are better heard in the original jurisdiction (Cambior); the arms' length relationship between head office and foreign subsidiary in which the Canadian-based head office cannot be held accountable for management decision of the subsidiary (Anvil Mining); and for substantive reasons related to the allegations against the company and its responsibilities under the law (Copper Mesa Mining Company) (Above Ground 2018).

However, recent cases have shown Canadian courts to be more willing to accept jurisdiction over these matters (MAC 2012, 52–3). Recent rulings in British Columbia (Nevsun Resources Ltd., Tahoe Resources), and Ontario (Hudbay) have allowed cases to proceed. Some of these cases address the important issue of whether companies are accountable for human rights violations by members of the security teams in the employ of their subsidiaries (Above Ground 2018; Quan 2017). The Nevsun case (on forced labour compelled by the state partner in the Bisha mine), on which the Supreme Court ruled in February 2020, has been viewed as opening the door to holding companies accountable in the court system for violations of international customary law, including of human rights. The Supreme Court did not provide guidance on those obligations, implying that common law jurisprudence will gradually define their contours (Hall 2020; Young and North 2014).

We must recall, however, that the cases considered by Canadian courts are civil cases and only involve potential monetary compensation to victims, should their claims be upheld. In no case are the substantive concerns of most anti-mining activists that drive conflicts, namely, a halt to mining activities, at play. In civil cases focusing on compensation for alleged wrongdoing, the standard for proving activist allegations is probably much higher than the CSR Counsellor's Office or the CORE, let alone, the "court" of public opinion. For example, the civil case against Copper Mesa (for compensation to victims of alleged brutality of the company's private security) failed because the plaintiffs did not succeed in providing the evidence, to the judge's satisfaction, to support a strong enough argument for "foreseeability and duty [of care]" (North and Young 2013, 98–9). It remains to be seen if lawyers representing local activists will have more success in making the legal and evidentiary argument in the cases currently before the courts.

Corporate legal responsibility for wrongdoing is further complicated by the definition and extent of complicity in alleged abuses. The cases

currently before the courts involve some of the most egregious allegations in which companies appear to have direct business relationships
with the actors directly accused of committing the violation (the state,
or security firms). It seems unlikely, however, that evidence will emerge
to tie management, by command, to many (or any) of these actions.
Instead, the landscape of allegations of human rights abuses, revealed
by a close read of the reports cited earlier, is one characterized by
ambiguous relationships between direct perpetrators and companies,
in which it is difficult to assess the *proportion of the company's involvement in the harm* (as the *Guiding Principles* would describe it). It will
be interesting to have the courts pronounce on these issues that really
go to the heart of the activist discourse about human rights violations,
namely, whether abuses committed independently by individuals and
organizations in a context that is partly or significantly structured by
the presence of the firm, but which cannot be directly tied to its management, are legally actionable.

Conclusion

We should remember that the social conflict generally occurs prior to
the occurrence of human rights violations. In this regard, a strategy to
prevent or reduce human rights violations should begin with reducing the severity or occurrence of social conflict between mining companies and communities. Host countries have the sovereign right to
determine where to permit mining and under what conditions.[3] Given
that starting point, does it matter if the home-country policy regime is
voluntary or legislated? This chapter has shown that both approaches
have some potential to address the problem but that neither is likely to
be a silver bullet. In this context, recourse to "hybrid accountability"
(Sagebien 2011, 118) that combines some form of government involvement with corporate self-regulation may be the most reasonable way
forward. Arguably, the CSR regime that constitutes an unofficial pillar of Canada's Americas Strategy has evolved along these lines from
being purely voluntary to weakly legislated.
    If our interest is in improving the state of affairs on the ground that
generate the emotionally charged contexts in which human rights
abuses typically occur, then there is some value in the voluntary
approaches that aim to change the processes and grievance mechanisms
through which corporations relate to stakeholders. The *UN Guiding
Principles'* emphasis on due diligence in human rights impact assessment is focused on transforming corporate governance to *preventatively*
consider risks to the infringement of human rights from the outset.

Indeed, this is where a company's CSR programme, and the government and Mining Association's promotion of CSR, has the potential to deliver some benefits and help to resolve or mitigate the severity of conflicts. There may be more potential for dialogue, mediation, and reaching a compromise that satisfies the parties to a conflict when the principal issues behind a conflict are distributional in nature (who gets what). The strength of this approach is that distributive issues are part of many, maybe most, conflicts even if they are not always *the* most important part (Arellano-Yanguas 2011; Haslam and Ary Tanimoune 2016; Amengual 2018).

If our interest is corporate accountability and compensation for the victims, *after the fact* of a violation, then the legislated option that establishes the investigative and mildly disciplinary function of the Ombudsperson's office, as well as access to the civil courts in Canada, offers advantages. Activists may well gain "moral victories" over companies and "public shaming" in a handful of cases that could clarify the human rights pitfalls associated with mining that are legally actionable and incentivize companies to take human rights due diligence seriously. After the Nevsun decision by the Supreme Court, the judicial road will play a much greater role in defining corporate social and human rights obligations. Yet, it must be admitted, that in cases of conflicts primarily based around a perceived incompatibility between modern mining and agricultural livelihoods that protagonists view as existential issues, neither process improvements internal to the firm, nor limited after-the-fact accountability in the corporate home country may be satisfactory to local actors.

In this regard, the Canadian government faces two highly imperfect home-country policy options that offer limited leverage over a complex and knotty problem, the real function of which may be political signalling to domestic constituencies. The added value of the disciplinary functions of the *Enhanced CSR Strategy* (2014), and the CORE (2018), will probably be found in the seriousness with which mining companies plan, execute, and report on their human rights impact assessments. It will be incumbent on the government to provide adequate guidance on these and related matters, while the effectiveness of these policy changes for people on the ground should undergo methodologically rigorous assessment. Let us be cognizant, however, that the home-country approach will not solve the problem of social conflicts and associated human rights abuses, nor adequately substitute for the development of host-country institutional capacity, citizen rights, and access to justice in Latin America – which could also be a priority for Canadian development policy.

NOTES

1 A notable exception is draft legislation prepared by the Canadian Network on Corporate Accountability, discussed in more detail below (see CNCA 2017).
2 Haslam, Ary Tanimoune, and Razeq's sample probably slightly over-represents the likelihood of being associated with a social conflict in 2012. See their methodological discussion (2018, 529–30).
3 It should be remembered that investment protection provisions found in bilateral investment treaties and free trade agreements are compensation regimes – that is to say, they require payment of compensation should the government change the rules of the game to the detriment of the firm (Haslam 2010; Dixon and Haslam 2016). While companies make exorbitant claims under this regime, awards are usually much less in value, and there is little evidence that governments don't pursue regulatory changes because of a fear of international arbitration. In fact, the regulatory and tax regimes for mining have undergone massive changes in Latin America since the early 2000s (Haslam and Heidrich 2016).

REFERENCES

Above Ground. 2018. "Transnational Lawsuits in Canada against Extractive Companies." (31 May). Ottawa: Above Ground. Accessed 2 November 2021. https://aboveground.ngo/transnational-lawsuits-in-canada-against-extractive-companies/.
Advisory Group. 2007. "National Roundtables on Corporate Social Responsibility (CSR) and the Canadian Extractive Industry in Developing Countries." Advisory Group Report (29 March).
Amengual, Matthew. 2018. "Buying Stability: The Distributive Outcomes of Firm Responses to Risk in the Bolivian Mining Industry." *World Development* 104: 31–45.
Arellano-Yanguas, Javier. 2011. "Aggravating the Resource Curse: Decentralization, Mining and Conflict in Peru." *The Journal of Development Studies* 47, no. 4: 617–38.
Bebbington, Anthony. 2012. "Social Conflict and Emergent Institutions: Hypotheses from Piura, Peru." In *Social Conflict, Economic Development and Extractive Industry: Evidence from South America*, edited by Anthony Bebbington, 67–88. Milton Park, UK: Routledge.
Bebbington, Anthony, Denise Humphreys Bebbington, Jeffrey Bury, Jeannet Lingan, Juan Pablo Muñoz, and Martin Scurrah. 2008. "Mining and Social Movements: Struggles over Livelihood and Rural Territorial Development in the Andes." *World Development* 36, no. 12: 2888–905.

Campbell, Bonnie. 2010. "Revisiting the Reform Process of African Mining Regimes." *Canadian Journal of Development Studies* 30, nos. 1–2: 197–217.

Canadian Centre for the Study of Resource Conflict (CCSRC). 2009. "Corporate Social Responsibility: Movements and Footprints of Canadian Mining and Exploration Firms in the Developing World." Revelstoke, BC: CCSRC.

Canadian Network on Corporate Accountability (CNCA). 2007. *Dirty Business, Dirty Practices: How the Federal Government Supports Canadian Mining, Oil and Gas Companies Abroad*. Ottawa: Halifax Initiative.

– 2016. "The Global Leadership in Business and Human Rights Act." Accessed 2 November 2021. http://cnca-rcrce.ca/campaigns-justice/ombudsperson/.

– 2020. "CNCA Public Letter to the CORE on Spring Consultation." Accessed 2 November 2021. http://cnca-rcrce.ca/wp-content/uploads/2020/07/CNCA-letter-to-CORE-re-spring-2020-consultations.pdf.

Canadian National Contact Point (NCP). 2014. "Initial Assessment of the Specific Instance under the OECD Guidelines for Multinational Enterprises Relating to Corriente Resources' Mirador Mine in Ecuador." (24 July). Accessed 2 November 2021. http://www.international.gc.ca/trade-agreements-accords-commerciaux/ncp-pcn/initial-assessment-evaluation-initiale.aspx?lang=eng.

Clark, Timothy D., and Liisa North. 2006. "Mining and Oil in Latin America: Lessons from the Past, Issues for the Future." In *Community Rights and Corporate Responsibility: Canadian Mining and Oil Companies in Latin America*, edited by Liisa North, Timothy D. Clark, and Viviana Patroni, 203–21. Toronto: Between the Lines Press.

Conde, Marta, and Philippe Le Billon. 2017. "Why Do Some Communities Resist Mining Projects While Others Do Not?" *Extractive Industries and Society* 4, no. 3: 681–97.

Coumans, Catherine. 2010. "Alternative Accountability Mechanisms and Mining: The Problems of Effective Impunity, Human Rights, and Agency." *Canadian Journal of Development Studies* 30, nos. 1–2: 27–48.

Dashwood, Hevina S. 2012. *The Rise of Corporate Social Responsibility: Mining and the Spread of Global Norms*. Cambridge: Cambridge University Press.

Dashwood, Hevina S. 2007. "Canadian Mining Companies and Corporate Social Responsibility: Weighing the Impact of Global Norms." *Canadian Journal of Political Science* 40, no. 1: 129–56.

Deneault, Alain, and William Sacher. 2012. *Imperial Canada Inc.: Legal Haven of Choice for the World's Mining Multinationals*. Vancouver: Talon Books.

Department of Foreign Affairs and International Trade (DFAIT). 2014. "Doing Business, the Canadian Way: A Strategy to Advance Corporate Social Responsibility in Canada's Extractive Sector Abroad." Accessed 2 November 2021. http://www.international.gc.ca/trade-agreements

-accords-commerciaux/topics-domaines/other-autre/csr-strat-rse.aspx
?lang=eng.

Dixon, Jay, and Paul A. Haslam. 2016. "Does the Quality of Investment
Protection Affect FDI Flows to Developing Countries? Evidence from Latin
America." *The World Economy* 39, no. 8: 1080–108.

Due Process Law Foundation (DPLF). 2014. "The Impact of Canadian Mining
in Latin America and Canada's Responsibility: Executive Summary of the
Report Submitted to the Inter-American Commission on Human Rights."
Working Group on Mining and Human Rights in Latin America, Due
Process of Law Foundation: Washington, DC. Accessed 2 November 2021.
http://www.dplf.org/sites/default/files/informe_canada_resumen
_ejecutivo.pdf.

Fasken Martineau DuMoulin LLP. 2012. *Corporate Social Responsibility
Accountability Report.* Ottawa: Fasken Martineau DuMoulin LLP.

Frynas, Jedrzej George, and Siân Stephens. 2015. "Political Corporate
Social Responsibility: Reviewing Theories and Setting New Agendas."
International Journal of Management Reviews 17: 483–509.

Garrod, J.Z., and Laura MacDonald. 2016. "Rethinking 'Canadian Mining
Imperialism' in Latin America." In *Mining in Latin America: Critical
Approaches to the New Extraction,* edited by Kalowatie Deonandan and
Michael L. Dougherty, 100–15. Milton Park, UK: Routledge.

Global Affairs Canada. 2019. Responsible Business Conduct Abroad –
Questions and Answers. Accessed 2 November 2021. https://www
.international.gc.ca/trade-agreements-accords-commerciaux/topics
-domaines/other-autre/faq.aspx?lang=eng.

Gratton, Pierre. 2017. "Statement to the Subcommittee." Subcommittee on
International Human Rights. Committee Meeting #75, 42nd Parliament,
1st Session. Ottawa: House of Commons. Accessed 2 November 2021.
https://openparliament.ca/committees/international-human-rights
/42-1/75/pierre-gratton-12/?page=1.

Hall, Matthew. 2020. "Liability: How a New Court Ruling Could Put
Canadian Miners in the Dock." *Mining Technology* (17 June). Accessed
2 November 2021. https://www.mining-technology.com/features
/liability-how-a-new-court-ruling-could-put-canadian-miners-in-the-dock/.

Haslam, Paul A. 2010. "The Evolution of the Foreign Direct Investment
Regime in the Americas." *Third World Quarterly* 31, no. 7: 1181–203.

– 2018. "Beyond Voluntary: State-Firm Bargaining over Corporate Social
Responsibilities in Mining." *Review of International Political Economy* 25,
no. 3: 418–40.

– 2021. "The Micro-Politics of Corporate Responsibility: How Companies
Shape Protest in Communities Affected by Mining." *World Development*
139: 105322.

Haslam, Paul A., and Pablo Heidrich. 2016. "From Neoliberalism to Resource Nationalism: States, Firms and Development." In *The Political Economy of Natural Resources and Development: From Neoliberalism to Resource Nationalism*, edited by Paul A. Haslam and Pablo Heidrich, 1–32. Milton Park, UK: Routledge.

Haslam, Paul Alexander, and Nasser Ary Tanimoune. 2016. "The Determinants of Social Conflict in the Latin American Mining Sector: New Evidence with Quantitative Data." *World Development* 79: 401–19.

Haslam, Paul Alexander, Nasser Ary Tanimoune, and Zarlasht Razeq. 2018. "Do Canadian Mining Firms Behave Worse Than Other Companies? Quantitative Evidence from Latin America." Canadian Journal of Political Science 51, no. 3: 521–51.

Haufler, Virgina. 2001. *A Public Role for the Private Sector: Industry Self-Regulation in a Global Economy*. Washington, DC: Carnegie Endowment for Peace.

Henisz, Witold, Sinziana Dorobantu and Lite J. Nartey. 2014. "Spinning Gold: The Financial Returns to Stakeholder Engagement." *Strategic Management Journal* 35: 1727–48.

Humphreys Bebbington, Denise, and Anthony Bebbington. 2010. "Extraction, Territory, and Inequalities: Gas in the Bolivian Chaco." *Canadian Journal of Development Studies* 30, nos. 1–2: 259–80.

Imai, Shin. 2017. "The 'Canada Brand': Violence and Canadian Mining Companies in Latin America." Research Paper No. 17. Legal Studies Research Paper Series, 13:4. Osgoode Hall Law School.

Jaskoski, Maiah. 2014. "Environmental Licensing and Conflict in Peru's Mining Sector: A Path-Dependent Analysis." *World Development* 64: 873–83.

Keenan, Karyn. 2013. "Desperately Seeking Sanction: Canadian Extractive Companies and Their Public Partners." Canadian *Journal of Development Studies* 34, no. 1: 111–21.

Kolk, Ans, Rob van Tulder, and Carlijn Welters. 1999. "International Codes of Conduct and Corporate Social Responsibility: Can Transnational Corporations Regulate Themselves?" *Transnational Corporations* 8, no. 1: 143–80.

Lipsett, Lloyd, Michelle Hohn, and Ian Thompson. 2012. "Recommendations of the National Roundtables on Corporate Social Responsibility and the Canadian Extractive Industry in Developing Countries. Current Actions, Stakeholder Options and Emerging Issues." Report for the Mining Association of Canada.

Martinez, Carla, and Daniel M. Franks. 2014. "Does Mining Company-Sponsored Community Development Influence Social Licence to Operate? Evidence from Private and State-Owned Companies in Chile." *Impact Assessment and Project Appraisal* 32, no. 4: 294–303.

Munson, James. 2017. "Fulfill Mining Ombudsperson Promise, Academics Tell Liberals." iPolitics (14 March). Accessed 2 November 2021. https://ipolitics.ca/2017/03/14/fulfill-mining-ombudsperson-promise-academics-tell-liberals/.

Natural Resources Canada (NRCAN). 2013. "Strategic Outreach and Partnerships Division." Corporate Social Responsibility Research Initiative. Final Report. Ottawa: Government of Canada.

North, Liisa, and Laura Young. 2013. "Generating Rights for Communities Harmed by Mining: Legal and Other Action." *Canadian Journal of Development Studies* 34, no. 1: 96–100.

Ponce, Aldo, and Cynthia McClintock. 2014. "The Explosive Combination of Inefficient Local Bureaucracies and Mining Production: Evidence from Localized Social Protests in Peru." *Latin American Politics & Society* 56, no. 3: 118–39.

Quan, Douglas. 2017. "'New Era': Canadian Mining Industry Closely Watching Three Civil Cases Alleging Human Rights Abuses." *National Post* (27 November). Accessed 2 November 2021. http://nationalpost.com/news/canada/new-era-mining-industry-closely-watching-three-civil-cases-alleging-human-rights-abuses.

Ruggie, John Gerard, and John F. Sherman III. 2017. "The Concept of 'Due Diligence' in the UN Guiding Principles on Business and Human Rights: A Reply to Jonathan Bonnitcha and Robert McCorquodale." *The European Journal of International Law* 28, no. 3: 921–30.

Sagebien, Julia, and Nicole Marie Lindsay. 2011. "Introduction: Companies and the Company They Keep: CSR in a 'Social and Environmental Value Governance Ecosystems' Context." In *Governance Ecosystems: CSR in the Latin American Mining Sector,* edited by Julia Sagebien and Nicole Marie Lindsay, 1–30. Houndmills, UK: Palgrave Macmillan.

Sethi, S. Prakash, and Olga Emelianova. 2006. "A Failed Strategy of Using Voluntary Codes of Conduct by the Global Mining Industry." *Corporate Governance* 6, no. 3: 226–38.

Speigel, Samuel J. 2012. "Governance Institutions, Resource Rights Regimes, and the Informal Mining Sector: Regulatory Complexities in Indonesia." *World Development* 40, no. 1: 189–205.

UN Working Group. 2018. "Report of the Working Group on the Issue of Human Rights and Transnational Corporations and other Business Enterprises on its Mission to Canada." Human Rights Council. Document A/HRC/38/40/Add.1. Accessed 2 November 2021. http://ap.ohchr.org/documents/dpage_e.aspx?si=A/HRC/38/48/Add.1.

van Tulder, Rob, and Ans Kolk. 2001. "Multinationality and Corporate Ethics: Codes of Conduct in the Sporting Goods Industry." *Journal of International Business Studies* 32, no. 2: 267–83.

Veltmeyer, Henry. 2013. "The Political Economy of Natural Resource Extraction: A New Model or Extractive Imperialism." *Canadian Journal of Development Studies* 34, no. 1: 79–95.

Walter, Mariana, and Joan Martinez-Alier. 2010. "How to Be Heard When Nobody Wants to Listen: Community Action against Mining in Argentina." *Canadian Journal of Development Studies* 30, nos. 1–2: 281–301.

Webb, Kernaghan. 2012. "Multi-level Corporate Responsibility and the Mining Sector: Learning from the Canadian Experience in Latin America." *Business and Politics* 14, no. 3: 1–42.

11 Conclusions

PABLO HEIDRICH AND LAURA MACDONALD

The chapters in this volume have documented the complex and multifaceted character of Canada's engagement in the Americas, seeking to contextualize what Canada's Americas Policy has been in the early twenty-first century. We have purposely investigated beyond the events and issues that have marked Canada's interaction with Latin America to provide better understanding of these as part of long-standing trends in foreign policy. In this final chapter we outline some of these ongoing challenges and opportunities that Canada faces in building a stronger and deeper relationship with the southern part of the hemisphere in the future. While starting from concrete, current by 2020, issues, we also provide insights as to what Canada will likely be facing in the near-term and the next decade.

NAFTA Renegotiations, Relations with Mexico, and Ongoing Trade Negotiations

Recent negotiations to amend NAFTA and eventually give birth to a new treaty, the USMCA, have brought for Canada, among other things, a meaningful collision between its Latin America policy and its North America policy. Those two policies, focused on distinct goals, instruments, and, above all, very different Canadian perspectives on who their interlocutors are, became spectacularly (or noisily at least) crushed into one as Mexico and the United States reached a bilateral agreement in August 2018, leaving Canada alone to settle on the roughly same terms with Washington later on. Much had been said before this point in North American capitals about the likely possibility of the Canadian government "throwing Mexico under the bus," but as González Haces clearly explains in her chapter of this book, that was not a possibility given the position Mexico had both, versus United States and versus

Canada, as it hosts significant investments of both and could be credibly undermined by the US border security policies.

Canada's lessons from this recent encounter with Mexico, its most significant partner in Latin America, is that its relatively weak understanding of Mexican politics, leverages, vulnerabilities, and capabilities as well as Canada's inherent sense of superiority over Latin American developing nations proved extremely expensive for its core economic and political interests back home. That shows the additional cost Canada has incurred when engaging more deeply with a neighbouring Latin American country such as Mexico without having developed a more sophisticated capacity to both engage Mexico directly and also address its growing importance for Canada's most important economic and political partner, the United States.

These lessons may be relevant in the future as Canada currently engages in trade negotiations with MERCOSUR (the trading bloc that include Brazil, Argentina, Uruguay, and Paraguay) and the Pacific Alliance (the newer trade bloc whose members include Mexico, Colombia, Peru, and Chile). Negotiating with these blocs has proven challenging, particularly with MERCOSUR, where Brazil's far-right leader, Jair Bolsonaro, clashed with Alberto Fernández, recently elected in Argentina, and built close relations with Donald Trump in Washington. Although Brazil is South America's biggest economy, Canada, which has been unable to deepen its relationship with Brazil even under more moderate leadership, faces today in Bolsonaro's controversial presidency a partner eager to sign free trade agreements but widely rejected by global and Canadian civil society. The dilemma for a free trade–seeking Canada now is how to square its economic goals with Brazil, by far the most important economy in the region, while negotiating with a government it openly dislikes. The cost of not advancing negotiations is an ever-closer relationship of Brazil with the United States, where Canada is once again sidelined. The benefit would be in maintaining a more coherent longer-term behaviour, demonstrating that serious concerns about human rights and democracy can, in Canada's case, overwhelm its short-term commercial interests.

In the context of a severely affected global economy due to the consequences from the COVID-19 pandemic, Canada can do well to leverage a more balanced approach between trade and human rights or democracy perspectives. Its so far intermediate results in containing the pandemic while ratcheting up extraordinary levels of government expenditures will necessarily generate an economy less trade driven and more debt financed, and one can also show better social outcomes than those in Latin America and even the United States. As this strategy is hard to replicate in the Global South, the divergence should be kept in mind by Canadian diplomats, NGO leaders, and business entrepreneurs expecting similar policies in the future Latin America.

Growing Impact of Mining in Canada's Latin American Engagement and NGOs

The use of Canadian diplomacy to support mining interests while simultaneously providing public funds to support Canadian NGOs with discourses critical of Canadian mining projects abroad is a domestically created incoherence with large implications in the relationship with Latin America. While such incoherence can be sold as a demonstration of pluralism (as done by successive Conservative and Liberal governments in Canada), it cannot escape the correlated inability of Canada to create appropriate institutions to manage the growing "mining-ness" of Canadian economic engagement with Latin America (and with other parts of the developing world).

In this volume, Moore's chapter clearly enumerates the costs for Canada in terms of social and national reputation, its heavy investment of diplomatic resources in defence of Canadian mining firms, and the probable meagre gains for average Canadian citizens. Her claim is that both Conservative and Liberal governments have been following a short-term approach to privilege the interest of Canadian firms and their quarterly profits from social protests, domestic policy changes in the countries where they operate, without regard to how Canada's image has become one widely disliked and even worse, seen as studiously hypocritical in matters of human rights, political sovereignty, and economic development.

Haslam seeks pragmatic policy alternatives to reduce the reputational and political risks for Canada from the involvement of some of its firms in human rights abuses. In his view, the alternatives are roughly two: more voluntary pressures on firms via support of CSR initiatives and public adherence to UN human rights principles, and a compulsory one through Canadian legislation on behaviour abroad and a public Ombudsperson to investigate firms' behaviours. According to Haslam, Canadian government policy has adhered to the former and, under pressure from NGOs, taken some timid steps in the direction of the latter. Nonetheless, even if acting in combination, both approaches would be able to do little since the underlying causes of human rights' violations are beyond Canada's diplomatic capacity to influence in view of the author.

The dilemma here is again not resolved and given the increasing centrality of mining in the relationship of Canada with Latin America, this problem remains, gaining relevance over time. This is partly an issue of proportionality; as Canada becomes increasingly specialized in this industry (in part because it is losing competitiveness in many others), the proportion of "mining-ness" in the Canada–Latin America relationship grows. Secondly, while Canada remains anchored in protecting the short-term interests of its mining firms, most Latin American countries are evolving in different directions regarding this industry, some outlawing it altogether, such as El Salvador and Costa Rica, others taking a slow road to approving more projects, such as Colombia and Panama, while others engage in further deregulation to facilitate more investments, such as Brazil, Venezuela, and Argentina. In the face of this diverse context, the "structurally conservative" position of the Canadian government regarding its mining investments abroad ends up being increasingly challenged, not only by NGOs and other human rights advocates abroad and at home but also challenged in terms of effectiveness, and as such becomes an ever-heavier burden for Canada on the overall relationship with Latin America.

The chapters by both Moore and Deonandan and Schmuland demonstrate together the depth of such incoherence on the part of the Government of Canada and its politicians. Deonandan and Schmuland explain how Canadian NGOs, often financed and therefore dependent on of public funds handed out by the Government of Canada's international development budget, needed to adapt to changing policy agendas in Latin America from Harper's conservative government and then Trudeau's liberal government. At risk of losing that precious funding to do their activities abroad, Canadian NGOs have had to moderate discourses or change activities regardless of the situations they observed in Latin America during the 2006–15 Harper governments, to later refocus on gender issues to a significant degree to toe the line of the Trudeau administration, again regardless of whether the NGOs themselves saw this as the most relevant development issue in the region or not in the current years.

Democracy Promotion and Security Issues

The distance between Canada and this region of its interest, Latin America, is made clear over and over again, as Shamsie demonstrates in her analysis of the case of Haiti, a country of outsize significance for Canada in the region as the only French-speaking nation in Latin America and one which has had a substantial flow of immigrants and refugees to Canada for decades. More consistent interest from Canada in Haiti over the years than in other Latin American countries have, in the view of Shamsie, not brought to the former greater clarity on how to support democratic transitions, such as the case she describes of the 2015 Haitian election. Instead, Canada has built over time a preference to act in coalition with other foreign actors, such as the United States and the European Union, to endorse or reject election results and concentrate its election observation aid in just one NGO also supported by these other Northern actors in Haiti. That rather counterintuitive result, a country as experienced with Haiti as Canada, instead jump on the bandwagon with other foreign actors keeping a line of its own, probably expresses Canadian policymakers' still insecure understanding of developing countries' realities. It also demonstrates again the lack of interest by Canada to learn from its own interactions and the societies or nations it seeks to influence in the South.

The lessons from involvement in Haiti are now evident in Canada's reaction to the deterioration of democratic norms in Venezuela, where that government has become ever more authoritarian under the presidency of Nicolás Maduro. While the Harper government had been unambiguous in its condemnation of the regime in Caracas, it frequently tainted its statements with criticisms of the economic policies of the Venezuelan version of socialism. As such, the Canadian position seemed more ideologically driven than really concerned with respect for democracy and human rights. Under the Trudeau government, Canada took a very long time to do anything, involving itself in the Lima Group that included most Latin American countries critical of Maduro's authoritarianism only after 2017. Once there, it simply imitated the positions being proposed by the harshest critics of the Venezuelan government in the region. That risk-reducing reflex effectively diminished again – just as in the Haitian case before – the credibility of Canada as a principled advocate of democratic and human rights.

The possibilities for a peaceful transition from political oppression and economic collapse in Venezuela, which have driven over 5 million people abroad as immigrants and refugees mostly to the rest of South America, have been repeatedly lost in the last years. In part by the clumsy Lima Group, but also by the lack of coherence and good strategies of the Venezuelan opposition. Canada has shown in this extraordinary crisis next to no leadership, little knowledge of how Venezuelan and even South American regional politics function, while making timid gestures of welcome to a few thousand Venezuelan refugees. Overall, a drop in the bucket for what has been the largest and swiftest national crisis in the region in the last decades.

The chapter by Rodríguez in this volume shows that Canadian economic interests also have implications for the country's foreign and security policies towards the region. Canadian foreign policy has often been dominated by a liberal approach to international security that views conflicts as stemming from underlying social and economic causes. This liberal tendency has resulted in some important contributions in the region, including Canada's support for peacebuilding efforts in Central America in the 1980s, as well as humanitarian assistance to Haiti. The increasing dominance of the mining industry in Canadian policymakers' approach to the region has, however, also undermined these liberal tendencies, particularly under the Harper government. The Trudeau government has adopted a softer rhetoric, with a strong commitment to themes of liberal internationalism and feminist foreign policy. Nonetheless, mining and other business interests remain powerful (as the case of the government's unwillingness to permit SNC-Lavalin to be prosecuted for acts of corruption carried out abroad). More worryingly still, is the continuously lax attitude towards human rights abuses by Canadian firms under the current government and its reluctance to make an Ombudsman position actually effective to monitor and eventually investigate violations.

The longer-term trend is one where Canada will continue to tag along with a Western model of extractive capitalism, now dressed up with inclusivity or progressive politics props (thus, the emphasis on feminism and Indigenous peoples) to reduce conflicts and resistance in Latin America or other parts of the South. As importantly, such discourses help Canada to define its diplomacy, NGOs and companies as different from Chinese or even Russian extractive ventures in Latin America and elsewhere. While the United States was absent from such rebranding exercises during the Trump presidency, opting instead for a mix of diplomatic inaction and bilateral direct threats (such as in the case of Venezuela), a return to a classical neoliberal tone under the Joe Biden presidency would provide Canada once again with the perfect leader to follow its own interests in Latin America's extractive landscapes.

The Wider Significance of Latin America for Canada's Global Engagement

As the depth and complexity of the relationship has grown, increasing possibilities have arisen for Canada (and for Latin America) for mutual benefits but also difficulties. Canadian policy towards Latin America has historically been informed by three main factors: its economic interests in the region, its social connections in the form of churches and NGOs, and its own domestic politics guiding foreign-policy discourses. On all three counts, this non-Commonwealth-nor-Francophonie part of the developing world has been an arena for the Canadian government and for Canada as a society to learn to act and project itself independently in the wider world in the later parts of the twentieth century and early twenty-first century. Since Canada shares some political, economic, and social conditions with Latin America (especially compared with other parts of the world, such as Russia, Asia, or Africa), this region has been a testing ground for several initiatives. This include Canada's decision in the 1970s to receive political refugees in larger numbers than before from Chile, and later from Central America and Colombia in the 1980s and 1990s and its partnership efforts that combined NGOs' solidarity efforts with government policy goals.

As Foster and later Deonandan and Schmuland detail in their respective chapters, NGOs and civil society actors in Canada have approached scenarios of political violence in Latin America from a prism of human rights and international solidarity from Canada. They achieved significant success in encouraging government actors from Ottawa to integrate these initiatives into Canadian foreign policy in the 1970s and 1980s but then became subject to regulation via financing and policy discourse regarding what these social agents from Canada could do in the region in the early twenty-first century. In other words, Canadian policy towards Latin America has had a social basis in this country, a fact often forgotten in the literature on relations between Northern and Southern parts of the world, but the role of the Canadian state has remained preponderant once those initiatives are integrated into the wider fold of foreign policy, often under the "development" subheading.

Besides the Canadian state–society relations demonstrated in the chapters here, a quality of learning for Canada to act abroad has taken place in Latin America. That could prove crucial today as the world becomes (again) more Asian-centric and Canada finds itself in frequent difficulties to establish stable and effective relations beyond its North Atlantic comfort zone. The current Trudeau's administration problems in dealing with China, Saudi Arabia, Russia, and India come to mind in this respect. To counter further diplomatic fiascos, Canada could well invest further into creating stronger diplomatic linkages with countries in Latin America, this time informed by deeper knowledge of their partners' politics and with more respect for their different opinions and priorities.

Not many shortcuts appear available to Canada anymore. It has traditionally leveraged its privileged participation in multilateral organizations such as NATO, OECD, Commonwealth, and Francophonie to project a disproportionate influence on UN institutions and regional development banks and from there, impact developing countries' policies in areas of its own economic interest (mining, banking, liberalized government procurement, etc.) or policy preference (R2P, human rights legislations, minority protections, etc.). It has been what Canadian foreign-policy experts have self-satisfactorily labelled as "punching above its weight." However, its recent experience of defeats at the UN Security Council elections (under both the Harper and Trudeau governments) and its loss of influence in the Inter-American Development Bank indicates that such a privileged position might not produce such significant benefits in the future. Furthermore, the rapidly changing architecture of global economic and political competition, with China, Russia, India, Turkey, and other rapidly growing but authoritarian and powerful nations from the Global South challenging the hegemony of the industrialized West, sets Canada into a rather automatic and not overly comfortable place of having to follow, most likely, erratic US leadership of that group.

To gain any degrees of freedom, Canada will have to double down on its efforts to build stronger bilateral and, from there, stronger regional relations beyond the West. Latin America is for the multiple reasons presented in this volume the place where Canada might have its best chances, but even there this will still require significant work beyond what it has committed so far. We hope that the research presented in this volume can be a fruitful reading to accompany that road, where it is hoped that both Canada and Latin America will create more positive and balanced relations.

Contributors

Kalowatie Deonandan, Political Studies, University of Saskatchewan

John W. Foster, International Justice, Human Justice (lecturer), University of Regina

Christina Gabriel, Political Science, Carleton University

María Teresa Gutiérrez Haces, Institute of Economic Research, Universidad Nacional Autónoma de México

Paul Haslam, International Development and Global Studies, University of Ottawa

Pablo Heidrich, Global and International Studies, Carleton University

Laura Macdonald, Political Science, Carleton University

Asa McKercher, History, Royal Military College of Canada

Jen Moore, Global Economy Program, Institute for Policy Studies, United States

Federmán Rodríguez, Political Science, Government and International Relations, Universidad del Rosario, Colombia

Toveli Schmuland, Faculty of Law (student), University of Alberta

Yasmine Shamsie, Political Science, Wilfrid Laurier University

Index

Note: Figures and tables are indicated by page numbers in *italics*

42–4; globalization and, 41; during "golden age" of Canadian foreign policy, 34–40; Harper government and, vii–viii, 3–4, 13–17, 24n1, 31, 48 (*see also* Americas Strategy (Harper government)); Indigenous peoples and, 6, 49; Justin Trudeau government and, viii, 5, 18–20, 24n2, 48–9, 82, 99–104, 158, 280, 281; liberal internationalist approach to, 12, 18, 22–3; mental maps and, ix, 30–1, 36–7, 39–40, 41, 42, 45, 47–9; Mulroney government and, 46–7; multilateralism and regional organizations, 7–8, 9, 10–11, 15–17, 18, 33–4, 36, 38–9, 42; neoliberalism and, 10, 12; northern *vs.* western hemispheric orientations, 29, 30, 34; Pearson and, 34, 35, 39–40; Pierre Trudeau government and, 8–9, 40–4; realist approach to, 21–2; theoretical approaches to, 20–4; trade, 15, 97, 199, 220n3, 276–7 (*see also* economic diplomacy; economic security; free trade agreements); trade balances with Mexico and US, 200, 221n4; US and, 4, 8, 9, 11, 13–14, 29, 33, 41, 197. *See also* Americas Strategy (Harper government); Canadian foreign and security policies (CFSP); civil society organizations (CSOs); democracy promotion; development assistance; foreign direct investment (FDI); Haiti, 2015 electoral crisis; Latin American Working Group (LAWG); Mexico–Canada relations; migration; mining sector; United States–Mexico–Canada Agreement (USMCA); *specific prime ministers*
Canada Latin America Resource

Centre (CLARC), 68, 76
Canada Not-for-Profit Corporations Act (NFP Act), 160, 165–6
Canada Revenue Agency (CRA), 164, 167–9, 187
Canada–United States Free Trade Agreement (CUSFTA), 4, 47, 72. *See also* North American Free Trade Agreement (NAFTA)
Canada–US–Mexico Agreement (CUSMA). *See* United States–Mexico–Canada Agreement (USMCA)
Canada Without Poverty (CWP), 167–8, 169, 170
Canadian Association for Latin America (CALA), 45–6
Canadian Association of Latin American and Caribbean Studies (CALACS; formerly Canadian Association of Latin American Studies (CALAS)), 46
Canadian Association of Petroleum Producers (CAPP), 207, 208
Canadian Centre for the Study of Resource Conflict (CCSRC), 254
Canadian Chamber of Commerce (CANCHAM), 213
Canadian Charter of Rights and Freedoms, 149, 168, 171
Canadian Commercial Corporation, 9
Canadian Conference of Catholic Bishops, 63–4. *See also* Catholic Church
Canadian Council for International Cooperation (CCIC), 165, 172, 180
Canadian Council for Refugees, 102, 149
Canadian Council of Chief Executives (CCCE), 150
Canadian Council of Churches, 102, 149

www.ingramcontent.com/pod-product-compliance
Lightning Source LLC
Chambersburg PA
CBHW030237030426
42336CB00009B/141